The political economy of soil erosion in developing countries

Piers Blaikie

Longman
London and New York

Longman Group Limited
Longman House, Burnt Mill, Harlow
Essex CM20 2JE, England
Associated companies throughout the world

Published in the United States of America
by Longman Inc., New York

First published 1985

British Library Cataloguing in Publication Data
Blaikie, Piers M.
  The political economy of soil erosion in
  developing countries.—(Longman development
  studies)
  ß Agriculture—Economic aspects—Developing
  countries.   2. Soil erosion—Developing
  countries   3. Agriculture and politics—
  Developing countries
  I. Title
  333.76'16   HD1417

ISBN 0-582-30089-4

Library of Congress Cataloging in Publication Data
Blaikie, Piers M.
  The political economy of soil erosion in developing
  countries.

  (Longman development studies)   SB 29619 £6.95. 7·86
  Bibliography: p.
  Includes index.
  1. Soil conservation—Government policy—Developing
  countries.   2. Land use—Government policy—Developing
  countries.   3. Soil erosion—Developing countries.
  I. Title.   II. Series
  D1131.B55   1985   333.76'13   84-879
  ISBN 0-582-30089-4

Set in 10/11 Plantin
Printed in Singapore by Selector Printing Co (Pte) Ltd

To Sally

Contents

Acknowledgements

The writing of books seems usually to be a very private and individual activity in which an author labours alone over a typewriter secluded from normal social intercourse. While my family and some of my colleagues will have suffered from this aspect of my authorship, I have greatly benefited from their contribution to the social and cooperative aspect of book writing.

The School of Development Studies at the University of East Anglia, for all its vicissitudes, still remains a most stimulating and friendly place to work. Although the recent cuts in spending on university education have taken their toll in the School in terms of innovativeness and time for research, many of my friends there have taken interest and trouble in my work and I owe them a debt of thanks. Steve Biggs, David Gibbon, Adam Pain and Mike Stocking read parts or all of the manuscript and gave valuable advice. Randall Baker, with whom I originally set out to write this book, has remained a close adviser, and some of his preliminary work in the book has been used, particularly in Chapter 4, where I hope that it is properly acknowledged to have contributed greatly to my understanding of the subject. Others such as John Harriss, David Seddon, Hugh Roberts and Ian Thomas have all contributed with references and advice. Two ex-students of Development Studies deserve special thanks. Sally Westwood worked part-time for three months on obtaining and summarising references. Her understanding of the book as it took shape was invaluable in providing me with a breadth of examples as well as theoretical material that would otherwise have been inaccessible to me. Sheila Chatting worked for three weeks during the Summer of 1982 and produced an astonishing output of concise and relevant digests of current research. Professor Denis Dwyer painstakingly read an earlier draft and made many valuable suggestions. Barbara Dewing drew the diagrams and maps, and Vera Ahrfeldt and Heather Latham edited the manuscript on the word processor. To all these, I owe many thanks and many interesting hours of cooperative work.

Lastly, the members of my family, Sally, Freya and Calum have put up with my mental and physical absences, and have always encouraged me in many different ways.

Piers Blaikie
Wacton, Norfolk.

Chapter 1

The issues addressed

1. The problem

Much of the debate about world-wide environmental deterioration is beset by uncertainty. One result of this is that there is a bewildering range of opinion about the causes, how far it has gone – and indeed about whether the whole issue is important at all. What is more, many of these contrasting views do not seem to address themselves to each other. One comment will be about a particular region of the world, from which general conclusions are drawn, but another will draw its illustrative material from another quite different region, and reach different conclusions. Also there seems to be a number of different ways of treating the whole problem, ideologically, politically and methodologically.

There are three major sources of this uncertainty. The first arises from the difficulty of obtaining accurate and widespread measurement of environmental deterioration through a long enough time period to indicate trends. Although there have been many recent improvements in measurement and monitoring, long established and reliable data sets are few and far between. Secondly, it is often difficult to single out the effect of humans on soil erosion and sedimentation rates, from other effects such as climatic change, and ongoing 'natural' erosion processes.

The last source of uncertainty provides the main focus of this book. It concerns the wide variety of ways in which environmental deterioration, and soil erosion in particular, can be viewed. Soil scientists, historians and social scientists all tend to have different approaches to the problem – what is important, what can be safely ignored, and the rules of discourse on the subject. Also there are widely differing political judgements involved, and many of these remain implicit and unexamined. This book aims to look at this source of uncertainty and make explicit the underlying assumptions which are made in the debate about soil erosion. However it is not a neutral book. It takes sides and argues a position because soil erosion *is* a political-economic issue, and even a position of so-called neutrality rests upon partisan assumptions.

Of course soil erosion is an *environmental* process as well. Within a geological perspective it goes on with or without human agencies. Even when a deterioration in soil structure and moisture retention levels occurs as a result of human activity, it may remain unrecognised or unimportant to people living in affected areas. It becomes a social issue when this deterioration (or removal of the soil) is recognised, and some sort of action taken. At this juncture it enters people's minds and becomes an object of thought. As long as this action (whether it be remedial, palliative or avoiding) is effective and does not cause a conflict of interests, soil erosion does not yet become a problem. Pastoralists can move to new pastures unchallenged, or farmers migrate to till new lands.

However, such adaptations, particularly during the last two centuries have usually brought about such a conflict. When this occurs, soil erosion also becomes a *political-economic* issue. The conflict of interests may be explicitly about soil erosion and conservation, or it may be politically expressed in other guises and connected to other issues such as a struggle for territory or over the amount or rate of appropriation of surpluses.

Because of the present-day recognition of soil erosion as a problem in many parts of the world, it is frequently disentangled from its political-economic context. Usually this recognition comes 'from above', either literally, in the case of a foreign advisor from a light aircraft, or by policy-makers, politicians or senior bureaucrats. Whoever recognises soil erosion, it is then publicised by those who have access to political power – and in a way which is at least not unfavourable to their interests, in so far as they may be involved.

After soil erosion has become a political-economic issue, and has been made an object of discussion and debate by those who have political power, then there arises the question of intervention, usually by the institutions of state. As this book argues, effective soil conservation involves itself in quite fundamental social change, sometimes involving people who live outside the affected area altogether. It is because the state becomes involved in soil conservation that soil erosion has *already* become a political-economic issue in the first place. So deeply embedded are the actions of human beings which lead to erosion, that intervention cannot avoid enmeshing itself in contradictions within society. Intervention therefore implies a mediation or a taking sides in conflicts of interest. The state is involved willy-nilly, and with it its functionaries – senior bureaucrats, politicians, foreign advisors, technicians, right down to the forest ranger and village headman. Intervention will usually affect (or try to affect) the livelihood of land-users in eroded areas, or at least where soil erosion is recognised to be a problem. This may involve land tenure systems; changes in the law, even in the constitution; rearrangements in pricing structures and credit arrangements; altering the foreign earnings capability of the nation, and so on. That is the problem and challenge.

2. What this book says

There are some leading opinions which claim that soil erosion, although perhaps widespread, is not important, and that 'induced innovations' by farmers, governments and private sector research and development institutions will cope. These opinions seem so diametrically opposed to many others which claim that erosion is widespread and serious that the problem arises over how to judge the issue (see sect. 2.2). Problems of reliable and comprehensive measurement in most lesser developed countries, the inadequacy of statistical models and the problem of distinguishing 'natural' from accelerated erosion, all make the likelihood of persuasive empirical proof seem remote. It is at least notionally possible to improve the level of empirical testing by implementing better monitoring systems, and developing an understanding of erosion processes and their statistical modelling, but the disagreements which make a settlement of the argument by empirical means unlikely are ideological ones. The principal issue here is a question of judgement over the extent to which land-users can mask or 'make good' soil degradation and erosion, and this depends upon their present and future access to chemical fertilisers, improved seeds, credit, government assistance in land erosion works, soil conserving crop rotations and tillage methods.

A number of serious doubts arise about whether induced innovations will be able to cope with soil erosion. First of all, it is not clear how poor farmers, often living in economically and physically marginal areas, will generate strong market signals to institutions of research and development in both state institutions and private concerns since such institutions tend to ignore the needs of poor and/or small farmers and environmentally fragile areas because the economic returns to research and development tend often to be small. It is also doubtful if there is the political will for governments to intervene in research as well as in effective assistance for these areas and people. Lastly, existing developments in agricultural technology are frequently not suitable or are inaccessible to small farmers and pastoralists in marginal areas. Political, social and economic barriers also exist to prevent affected rural populations from migrating to new, uneroded lands or from leaving the agricultural sector altogether. It is a matter not of what exists but of who commands what.

These considerations complete the vicious circle. Inequalities between the majority of the rural populations affected by soil erosion and other more powerful groups in access to adequate economic opportunities are both a result and a cause of soil erosion. In this sense soil erosion is a symptom of underdevelopment, and it reinforces that condition. Therefore to the rural poor at least, soil erosion is an important element in their poverty. There follows a simple typology of social contexts in which soil erosion occurs and it shows a great range of types of societies in lesser developed countries as well as in western and socialist countries. Both the processes of development and underdevelopment can lead to serious soil erosion (see sect. 2.3).

In an evaluation of existing conservation in lesser developed countries a distinction has to be made between techniques, programmes and policies (sect. 3.1). A brief review of the techniques of soil and pasture conservation follows (sect. 3.2), with some brief indication of the degree and type of social intervention that each technique requires.

While conservation techniques are the direct method of reducing or preventing soil erosion, programmes and policies involve implementing those techniques and therefore confront much more problematic political and economic issues. While techniques within pilot projects may have enjoyed some success, their extension and implementation over wide enough areas and numbers of people to have an appreciable effect in reducing erosion and improving food production have almost universally met with slow progress or failure. A review of policies of lesser developed countries and some other countries shows that only South Korea, the Republic of South Africa, and parts of the People's Republic of China can lay any claim, however qualified, to a successful national policy (sect. 3.3). This is not to deny the success of smaller projects which do not test the implementing capacity of government institutions nor involve themselves in large-scale politically sensitive issues.

Why do these policies usually fail? In Chapter 4 the starting point is to analyse different perceptions of the problem, especially of policy-makers, government officials and administrators. If it is recognised that 'social factors' are important in the problems faced by conservation policies, they apply not only to land-users but also to those who would intervene (see sect. 4.1). A 'colonial' or classic model of soil conservation is identified in which the problem of soil erosion is seen primarily as an environmental one, rather than a complex 'socio-environmental' problem, and therefore coercion and force applied by the colonial rulers upon the ruled can afford to ignore the social problems which led to soil erosion in the first place (see sect. 4.2). The second attribute of the colonial model is that it lays the blame on land-users themselves, and identifies them as lazy, ignorant, backward or irrational. Thirdly the model closely links the issues of overpopulation and soil erosion, and lastly it assumes that one of the major policy directions is to involve cultivators and pastoralists more strongly in the market economy. The economic and ideological structure of the colonial period goes a long way to explain this perception of the problem. For a number of reasons, elements of this colonial political-economic structure still persist, again both in the realm of economics as well as in that of ideas.

Many conservation policies are initiated, financed and staffed by foreign aid donors (sect. 4.3). The particular requirements of successful conservation programmes are just those which foreign aid projects are not very good at supplying. Programme inputs and outputs are diverse, diffuse, long term and difficult to quantify both conceptually and practically. Implementation is deeply involved in internal political issues; and at all stages of project design, implementation and evaluation, in-depth and critical political economic analysis and uncomfortable recommendations usually have to be made.

The state of the art of soil conservation has slowly evolved from the colonial model, although some of its characteristics remain in present-day policies. A mood of self-examination among policy-makers has been prevalent for some time, and a number of shortcomings have been identified. First, technical failures in soil conservation techniques have been seen to have been the cause of a number of problems in conservation policy. Second, many conservation techniques do not fit in with existing agricultural or pastoral practice, nor are they chosen or implemented with the participation of the land-users themselves. Lastly, many institutional weaknesses, particularly a lack of coordination between line agencies have also been identified (see sect. 4.4). These admissions of failure are useful, but they amount to a piecemeal improvement in the perception of the problem. Most of these reasons for failure are related to each other, and although they can prompt some remedial action (e.g. the stimulation of the flow of information between implementing agencies or the linkage of conservation with wider development efforts), they fail to make explicit more fundamental assumptions. These are that the causes of soil erosion may be outside the afflicted area altogether; that the state is not neutral and cannot necessarily solve soil erosion problems rationally or impartially; and that there are always winners and losers in erosion and conservation (see sect. 4.5)

A new approach is suggested in Chapter 5. There are considerable problems in combining the study of essential physical and social processes in the political economy of soil erosion which derive from the rather different epistemological and ideological characteristics of the natural and social sciences (see sect. 5.1). Such a study must include a 'place-based' analysis of soil erosion – where it actually occurs, where flooding and siltation caused by soil erosion in one place affects another, and where land-users have been spatially displaced to and from. It must also include and combine 'non-place-based' analysis of the relations of production under which land is used, the technology used and why, prices, taxes and so on.

Bringing these two analyses together (see section 5.2), a 'bottom-up' approach is outlined in which focus is first directed to the smallest unit of decision-making in the use of land (the family, the household, pastoral group), and the politics within the household (sexual politics, and the role of the 'head of the household'). Next, the household or small group's decision-making (over what crops to grow, where to pasture stock, where to collect fuel and fodder, etc.) is affected by a larger political organisation such as the village council, tribal chief or other representative of a local and territorially defined political institution. Finally, the focus falls on the way government and the administration effect land-use decisions through nationally enforced legal structures (particularly land tenure); pricing policies; taxation; allocation of national resources; and through more general but pervasive influences on class power and struggle – where power lies and how it is used (see sect. 5.3).

A review of how class interests in erosion and conservation are expressed follows (see sect. 5.4). Where there are people involved in soil erosion (directly and indirectly) and in conservation who do not have political power and cannot effectively express their economic interests through formal or informal channels, their responses tend to be individual, usually concerning private land-use decisions. The necessary cooperation with other land-users is usually difficult to achieve. They are often diverse, politically weak and divided, and the problems caused by erosion or conservation are frequently bound up in a variety of other conflicts of interest. Therefore concerted political action tends to be rare, and when it occurs it is usually ephemeral. Those groups or classes who tend to have access to and use the 'state apparatus' are seldom adversely affected by soil erosion, and what adverse effects they do experience can be avoided. Sometimes their power derives from the fact that they are surplus-producing farmers who export food to towns or abroad (thereby earning foreign exchange), and they can often prevent government 'interference' in their land-use and investment decisions on the farm. So ruling classes in many less developed countries are frequently not too concerned at the effects of soil erosion. Senior bureaucrats and policy-makers whose function it is to make and implement policies therefore often face lukewarm responses, and many embarrassing political contradictions.

New problems arise from this approach (see sect. 5.5). The analytical task is enormous, and action may be called for that stretches far beyond the individual's sphere of possible action and terms of reference. There are also problems of what some natural and social scientists expect as acceptable rules of discourse, proof and evidence. As an illustration of the new approach suggested, a comparison follows between soil conservation and family planning policies as examples of government-sponsored development activities (see sect. 5.6). They are similar in many respects. First, they both attempt to get individuals to change a most intimate and important part of their lives – of work and reproduction. Second, there are contradictions between private and public benefits. Third, there are technical reasons (in the design of contraceptive devices or soil conservation techniques) why adopters sometimes suffer. Fourth, land-use and family decisions are dependent on a mass of other variables which only change slowly and are not amenable to precise and rapid policy initiatives. And lastly, both programmes tend to be unpopular and politically embarrassing to government. There have been significant advances in the conceptualisation of rapid population growth and family planning which are relevant to soil erosion and conservation, particularly in the political economic analysis of why people have large families or contribute to soil erosion, and in whose interests and in the name of what dominant ideology the state intervenes.

The book then shifts attention away from the conceptualisation of the soil erosion problem and its attempts to intervene in the form of conservation policies, to the causes and implications of erosion itself. Chapter 6 presents a heuristic and formal model to explain how and why individual decision-makers (farmers, pastoralists, and other land-users) may cause soil erosion. It is based on a dynamic access model where income opportunities (with specified entry costs and rewards) are available to households who fulfil those entry costs. These income opportunities are both agricultural (e.g. different crops planted at different times, with different tillage techniques and other inputs) and non-agricultural. At the end of each season or year families accumulate, remain stable or disinvest, sometimes involving a gradual loss of control over production assets. This process is then 'laid over' a map of erosivity and erodibility, and the results of these decisions followed through in terms of erosion and conservation. Feedback effects upon the land-users themselves are analysed.

Chapter 7 examines the political-economic structures and processes which lie behind the model. Modes of surplus extraction from peasantries and pastoralists are most important in obliging land-users to take out of the soil, pastures and forests what they cannot afford to put back (see sect. 7.2). The integration of land-users in lesser developed countries into the world economy (see sect. 7.3) involved displacement and often confinement into a smaller land area; taxation to encourage them either to produce for the market or to work in the mines or plantations of settlers; and a general encouragement to do both by the provision of agricultural inputs and consumer items. Local histories, however, show very considerable variations. These processes continue today with fundamental implications for land use. Types of land-uses which were extensive suddenly started to cause environmental deterioration when confined to smaller areas; shortages of labour occurred at times of peak demand; the introduction of cash crops caused a reduction in food security, and often methods of tillage and planting which were much less conserving of soil than the multi-cropping and inter-cropping of food crops which it displaced. The precise relationship between land-users who both work their own land using their own implements as well as labouring for wages (e.g. in nearby large plantations or in towns or mines) is central to understanding the process of surplus extraction and for land-use. Three related processes of marginalisation, proletarianisation and incorporation are defined in section 7.4, and are illustrated by a number of extended case studies in central India, Niger, Nepal, Mexico and Zambia. The problem of the 'commons', and its relationship with private property is examined. Privately owned land (often arable land) coexists with commonly-held land (often pasture, and situated in steeper areas and on interfluves). Pressures upon both types of land due to differentiation among land-users and surplus extraction are examined, and the mutual feedback of environmental degradation between the two are discussed, using a case study in southern India. The discussion then shifts from a consideration of class relations in the use of land to particular areas which have a particular syndrome of symptoms – mountainousness, and/or low rainfall, and a history of 'semi-colonisation' where imperial powers could not subdue these areas militarily and instead came to an accommodation with their rulers. Subsequent developments led to population growth, a stagnant agricultural technology, outmigration with a remittance economy, and serious soil erosion (see sect. 7.6).

In Chapter 8, larger enterprises using land are studied, such as plantations, logging companies, large-scale ranches, state farms and so on. Their political power makes them more or less immune from interference from government (see sect. 7.2). The agricultural techniques used for commercial-scale agriculture and some of the reasons why increasing soil degradation and erosion do not necessarily bring about adaptations in their technology are reviewed (see sect. 7.3). The case of logging companies and their wholesale evasion of sound felling practice is a case in point (see section 7.3). The conclusion which is drawn in section 7.4 is that there is a strong tendency for conservation to occur only when erosion seriously affects the accumulation possibilities of the most powerful classes (including industrial capitalists where erosion may cause siltation of dams and damage to hydroelectric stations). The book ends with Chapter 9 entitled 'What now?' which is in itself a summary, conclusion and a discussion of implications for the future.

3. Scope

This book is about soil erosion in lesser developed countries. Since soil erosion also occurs in developed countries both capitalist and socialist, it might be asked why the book confines itself in this way. The answer is not that it is written from the viewpoint of a developed society telling a lesser developed one what should be done. The book concentrates on lesser developed countries because it is there that processes collectively called 'underdevelopment' are particularly acute, and one of them is environmental degradation. Thus environmental degradation is seen as a *result* of underdevelopment (of poverty, inequality and exploitation), a *symptom* of underdevelopment, and a *cause* of underdevelopment (contributing to a failure to produce, invest and improve productivity). For this reason soil erosion, in so far as it can be disentangled from the process of underdevelopment, is a particularly serious phenomenon. In terms of absolute deprivation, it can contribute to chronic food shortages, so-called natural disasters and hazards, drought, landslides, floods, and undermine an entire country's development effort, while in developed countries its impact is less immediate and pressing. Furthermore, since the lesser developed countries are predominantly situated in the tropics and sub-tropics, there are strong environmental reasons why soils and pastures tend to be much more fragile than in temperate climates (Young 1976; Grigg 1970: 190–205). Thus certain land-uses in the tropics can bring about very rapid and sometimes irreversible environmental damage. Whole landscapes can be transformed into an unproductive wilderness in a matter of a decade or little more. Putting both reasons for the potential seriousness of soil erosion in poor countries together, a third appears – many of the people of such countries are too poor and politically impotent to reverse the process. In any conservation issue, a stitch in time saves nine, and few people in lesser developed countries can afford nine stitches... or even one, in some case studies discussed in this book.

A second limitation on the scope of this book is that it excludes other important processes of environmental degradation in lesser developed countries, such as alkalinisation, salinisation, waterlogging, weed infestation, bush encroachment as well as industrial, and urban pollution (e.g. Eckholm 1982, addresses the whole range of these problems). The method used to analyse soil erosion and conservation can be extended to these other types of environmental degradation as well. This emphasis on soil erosion and degradation is dictated more by the breadth of subject matter of these other types of degradation and the relatively short length of the book. The main difference between soil erosion and alkalinisation, salinisation and waterlogging is that the latter are usually easier to measure, and the role of human activity is relatively easy to pinpoint.

Thirdly, the title of the book may imply that it is only soil erosion rather than soil degradation or pasture degradation which is being discussed. This is not the case. While a strict interpretation of soil erosion means the removal of inorganic soil grains (Kirkby & Morgan 1980: 7), a broader interpretation after Rauschkolb (1971) and Hudson (1971) is used in this book. Hudson (ibid: 41) says: 'If soil erosion is defined in the widest possible terms, it includes any degradation of the soil which reduces its ability to grow crops.' He includes in the definition of soil erosion 'fertility erosion' (loss of nutrients); 'puddle erosion' (physical breakdown of soil by unimpeded raindrops leading to the choking of soil by the washing of fine particles into the interstices of larger particles in the surface layers of soil); and 'vertical erosion' (washing of clay particles through gravels or sands to accumulate at lower depths in the soil profile). These three types of erosion do not necessarily imply physical soil removal. They are collectively called soil degradation by some authors and are usually, even closely, followed in time by soil erosion in the narrow sense. When organic matter in soil declines appreciably, moisture retention is reduced, so also reducing soil permeability, leading to an increase in overland flow, in turn causing physical removal of the soil. Also reduced organic content of the soil generally leads to reduced vegetative cover which increases the damaging effect of excessive solar heating and rainsplash. Thus degradation is closely allied to physical removal of the soil and the extent and timing of the latter will depend upon erodibility and erosivity. These latter terms refer to those factors which determine the ability of the soil to withstand erosion on the one hand, and the forces which act to erode soil on the other (e.g. the periodicity and intensity of rainfall, wind speed etc.). Although it is often important to distinguish between erosion in the narrow sense and degradation, the word 'erosion' as used in this book implies both.

The last aspect of the scope of the book is most difficult. If the approach to soil erosion is that it is not only an environmental process, but also a social and political-economic process, there must be some strong and important assumptions about how 'political economy' operates. There is a wide range of political economic theories (e.g. those of Adam Smith, David Ricardo, Thomas Malthus, or Karl Marx), and there are also other theories of human behaviour which would not claim to be political economy at all. Foremost among these is neo-classical economics, which some might claim is the theorised and quantified derivative of eighteenth and nineteenth century political economy, in which the 'political' and 'social' have been in effect abstracted out and marginalised. A political economic analysis of soil erosion must therefore identify its position and define its terms – and here lies the problem. A very brief separate exposition would amount to little more than reciting an article of faith, and would not allow any analytical defence. Any longer such exposition would run the risk of irrelevance. Therefore, as has been stated already (sect. 1.2) and is explained in more detail in sections 5.2 and 6.2, a 'bottom-up' approach to the problem is employed, which starts with land-users who make decisions on how soil, crops, livestock, grasses, trees and other materials are to be used, and then which works up to larger aggregates of land-users and their relations with others. In the course of this progress, the relevance of certain debates in political economy becomes plainer, and it is in this *ad hoc* fashion that they are introduced, and the stance of the author explained. This book tries to avoid using the issues of soil erosion and conservation as a basis for a wide-ranging polemic against the world economic system. There are ideological assumptions made in the course of argument, and these are labelled as clearly as possible. The immediate causes and results of soil erosion remain the focus of the book, and they are not used as sticks to beat a more comprehensive target.

Chapter 2

Is soil erosion really a problem?

1. Conflicting views

'Of course arable land in some places is going out of cultivation because of erosion and other destructive forces. But taken as a whole, the amount of arable land in the world is increasing year by year.' (Simon 1981: 81).

This view forcibly made by Simon claims that land is like any other resource which, far from being finite as for example, the Club of Rome world models assumed (Meadows et al. 1972: Mesarovic & Pestel 1976), actually becomes *cheaper* and more plentiful as technology finds and creates new resources and uses existing ones more efficiently. In this context, soil degradation exists but, taken as a whole, is unimportant. The more technically advanced agriculture becomes, the smaller is its dependency on natural endowments.

A similar opinion is expressed by Beckerman (1974: 239–40):

> As regards the physical limitations in food supplies, there seems to be general scientific agreement that this does not constitute the real constraint... the vague counter argument that more intensive cultivation will ruin the soil is hardly convincing in view of the fact that soil has been farmed with increasing intensity in Western Europe for about 2000 years and there is still no sign that it is exhausted. Most of the world's cultivable areas are, by comparison, either hardly touched or not yet touched at all. Nor does the need for more fertilisers for purposes of modern agricultural methods pose any problem, as there are ample supplies of phosphates and potash.

Although addressing a different problem, Ruttan (1982) along with Boserup (1981) both put forward the idea of 'induced innovation' as the means by which research and development of agricultural technology is directed to address appropriate problems both by the state and farmers themselves. Ruttan cites the contrasting examples of the USA and Japan. In the USA where labour was the scarce resource, it was the process of mechanisation first with animal and later tractor motive power that facilitated the expansion of agricultural production and productivity by increasing the area operated per worker: while in Japan where land was the scarce resource, it was progress in biological technologies leading to increased responses by varieties of paddy to higher levels of fertiliser application. In both cases a process of dynamic adjustment to changing relative factor prices was achieved, and the appropriate innovations induced successfully by these implied price signals. Ruttan believes that new agricultural technology and new institutions to develop it will be induced into existence to meet the challenge. Therefore he places emphasis on the importance of each country ensuring that factor prices reflect scarcities, thus ensuring that new technology will always develop in the most efficient market-led directions. However, he does not consider either the costs and returns to different groups in society nor different regions of the world. Furthermore the issue of environmental degradation as a result of induced innovation is hardly given any attention at all.

Boserup on the other hand gives explicit attention to environmental degradation, but in such a way as to suggest that in some cases soil erosion actually induces desirable agricultural innovations. In her most recent book (1981: 50) she gives examples where the destruction of topsoils in the reaches of a watershed, through population pressure leading to destabilising agricultural practices, had induced intensive agriculture in the valley floors which had been fertilised by the removal of topsoils further up-valley. She quotes examples of environmental deterioration in China for example which started to be acute between AD, 1500–1750 and which induced long distance transport of night-soil, labour intensive digging of river silts, widespread terracing, and recycling of residues and wastes. The ability of modern technology to cope with the problem of soil erosion is summarised thus: 'Growing populations may in part have destroyed more land than they improved, but it makes little sense to project past trends into the future, since we know more and more about methods of land preservation and are able by means of modern methods, to reclaim much land, which our ancestors have made sterile.' (Boserup 1981: 22)

Although these four authors address themselves to rather different problems, and do not appear to agree on a number of issues, their combined views on the importance of soil erosion are more or less consistent, and quite clear – its importance is exaggerated, agricultural technology initiated by both officials and farmers can respond satisfactorily, and the resources at disposal are massive and are constantly being created by technology itself. This view is in complete contrast to those who maintain that environmental deterioration is widespread and of critical importance. So contrasting is it that one may wonder whether its adherents are looking at the same thing at all: 'As the eighties unfold, humanity faces a worldwide shortage of productive cropland, acute land hunger in many countries, escalating prices for farmland almost everywhere...' (Brown 1981 in an article entitled 'Eroding the basis of civilisation').

Two more quotations of many hundreds serve to make the point that the distance between the two views is so large that they are probably not going to be reconciled by proof or disproof of the physical evidence.

> There is an environmental nightmare unfolding before our eyes... it is the result of the acts of millions of Ethiopians struggling for survival: scratching the surface of eroded land and eroding it further; cutting down trees for warmth, fuel and leaving the country denuded... (Eckholm 1976: 77).
> Effects of soil erosion in every degree of intensity can be found in Latin America, ranging from ravaged mountain slopes of the Andes to incipient dust storms in rangelands of the Argentine Republic. According to preliminary estimates, twelve of the southern countries have been damaged by severe erosion to the extent of about 70 million acres out of some 140 million acres of land now in crops... in six of these countries, Chile, Colombia, Ecuador, Honduras, Mexico and Venezuela, the estimated amount of land mined or severely damaged ranges from about 50 per cent to more than 60 per cent of the area appraised as arable (Bennett 1944: 63)

A study of a remarkable set of documents put together by USAID (the administration for foreign aid given by the United States) during the past few years entitled, *Environmental Profiles,* one for each country, shows in considerable detail the extent of soil degradation, alkalinisation, salinisation, waterlogging, and desertification world-wide. Almost every nation in central and Latin America, Africa and South East Asia seems to be suffering from environmental decline, and although the basis of quantification might be questioned, at least there is the framework of a sub-regional, quantified data base on environmental decline. Also references in this book to the extent of soil erosion are drawn from a wide variety of sources and go some way to corroborate the view that, whether it is important or not, environmental degradation is certainly happening. Thus we do not intend to provide evidence here of frightening rates of soil loss and desertification throughout the world. Eckholm's book (1976) does this more than adequately, and less recent accounts (e.g. Jacks and Whyte 1939: Hyams 1952) have also documented what has been seen as a widespread and serious problem. The view taken here is that one can produce quite easily a catechism of doom, supported with prodigious rates of soil loss over millions of hectares (and some examples are given in this book). However, attention is instead directed in this chapter to an ideological examination of the question of whether all this *matters,* and if so, how and to whom. The view that will be taken is that it does matter very much, although not quite in the way in which previous writers about environmental doom have envisaged. The rest of the book addresses itself to developing this approach.

2. How do we judge?

There are two interrelated issues about the rules of discourse on the subject. The first is that of proof. What constitutes proof, and is it possible? In other words, can Beckerman or Simon on the one side, or Eckholm or Lester Brown on the other (as examples of leading protagonists) be *proved* wrong? The second is related and is a question of ideology, and at a more applied level, one of politics–whether environmental deterioration matters depends not only on proof but also on one's view of social change and development.

Let us start with the problem of proof. The most pervasive characteristic of measurements of soil loss at different points throughout the world is their scantiness.

In a review of the present situation of soil degradation in ten countries of the Near East, Rafiq (1978: 9) states that: 'in most countries only qualitative information is available,' a view which Stocking (1978a) shares with reference to Africa. The UN Conference in Desertification (1977: 177) also agrees: 'Statistics are seldom in the right form, are hard to come by and even harder to believe let alone interpret.'

We can identify a number of interlocking and mutually reinforcing reasons why the situation prevails.

First, there are few resources in developing countries to measure soil loss. Quite sophisticated equipment and trained manpower is required to make reliable estimates. Furthermore measurement has usually to be made over a period of time. Compared with the measurement of alkalinisation and salinisation, whose samples of soil can be easily gathered at one point in time and the electrical conductivity of saturation extracts gauged, the measurement of soil loss is more demanding.

Second, the range of data required tends to be demanding, for example see Skidmore (1977) and Skidmore, Fisher, and Woodruff (1970) for the sophisticated data needed to assess wind erosion. Usually the physical data on soil loss has to be combined with a fairly detailed knowledge of land-use and its history, if the man-made contribution to erosion is to be reliably estimated over large areas (outside carefully laid-out experimental plots in research stations).

Then there is a group of more technical reasons which make statistical modelling and extrapolation more difficult, and inhibit attempts to overcome a shortage of resources for direct measurement by statistical estimation of soil losses. There are problems related to the complex and varying manner in which symptoms of soil erosion appear through time. Douglas (1981) summarised the time-dependent effects of soil eroding land-use practices in terms of changing soil erosion soil nutrient status, soil induration, weed invasion and crop yield, as well as upon the impact on rural incomes and the rural population. It is clear that the timing of these impacts varies enormously between cases, and that long time periods are required for some impacts to emerge. Furthermore, simple measurements of sediment at a point represent only the *net* result of all the processes going on upstream. The measurements do not identify the spatial nor temporal source of sediment. The sediment volume measured could have resulted from a long past phase of forest clearance in an upstream catchment – or a more recent event in a completely different location altogether.

In addition, the scale of measurement over which soil erosion occurs affects the measurement itself. As Gardner (1981) points out in relation to the FAO/UNEP World Assessment of Soil Degradation mapping exercise (FAO 1979: Riquier 1978, 1982), the universal soil loss equation has been misapplied in the estimation of soil erosion in quite large territorial units. 'In all but the most uniform of plains, the spatial anisotropy of soils, vegetation and topography is such that, at a scale of 1:5 million (that of the FAO/UNEP maps), a single meaningful measure of the parameters is not possible.' (Gardner 1981: 2) General problems of the use and misuse of the universal soil loss equation are discussed by Wischmeier (1976) and Stocking (1978a); these contribute to the difficulty and extrapolation particularly over large areas implied by the scale of the FAO map. The scale of measurement also affects the estimation of soil erosion because, as it becomes smaller, the problem of redistribution of eroded material *within* the areal unit becomes more serious. Erosion may be very considerable, but, because the area of net measurement of soil movement is large, sedimentation may occur within the same unit and therefore net erosion is estimated to be small (see also Blandford 1981).

Further, the relationship between erosion and crop or livestock yield have not been reliably modelled. There are separate models for erosion processes and others for crop growth processes, but the link between the two sets tends to be under-explored (Dumsday & Flinn 1977; Amos 1982: Ch. 2). For example erosion rates in a vertisol will have very much less of an impact upon yield than in a granite sand. One attempt to model only critical variables which determine yield has been made by Stocking and Pain (1983) where they concentrate upon the impact of erosive forces upon the availablity of minimum moisture levels.

Finally, the question of proof about soil erosion demands that natural and man-made or enhanced rates of soil loss can be distinguished. Partly this is a matter of time and related to the first problem discussed above. Only a marked increase in erosion rates, coupled with evidence of changes in agricultural practice and/or extent of agriculture, might in these cases provide sufficient evidence of accelerated erosion. It has been recognised how difficult it is to attribute accurately the effect of humans on soil erosion (Stocking 1978b: 130) because of the uncontrolled nature of many other crucial variables, such as climatic change. Therefore it is difficult to interpret such statistics as those of UNEP (1982), as well as others which calculate rates of topsoil loss per km^2, except perhaps in cases where the rate is extremely high and where there is other evidence of obvious changes in land use. Statistics quoted at length in UNEP 1982 have shown the amount of land converted into lower classes of production (relative to previous periods) by soil degradation and erosion. UNEP (ibid.) shows that all classes of cropland are predicted to increase in area up to the end of the millennium, but that large quantities of land (most certainly concentrated in marginal environments) will lose productive capacity. In the same report (UNEP 1982: 265), it is claimed that the total area being reclaimed by irrigation is probably about the same as is being abandoned through salinisation, alkalinisation and water-logging, and that these problems account for the loss of about 2–3 million ha of the world's best agricultural land each year. The assessment of the seriousness of soil erosion is frequently done on the basis of the percentage reduction in yields as well as, or instead of, objective physical measures of changes in properties of the soil (e.g. Rafiq 1978: 36). Tolerable soil losses inevitably involve political issues as well as technical ones (such as the natural rate of soil formation, or the 'T' factor as calculated by the USLE (Universal Soil Loss Equation) developed by the Soil Conservation Service of the United States. In any case there are considerable technical problems in assessing the 'T' factor which is defined as the maximum rate of annual soil erosion which would permit high-level crop production to be maintained indefinitely (Schertz 1983). Cook (1982) argues that the measure is political and moral without much rational basis and credibility.

At this point the discussion shifts to the question of ideology in assessing the importance of soil erosion. The central issue here is the assumptions about *future* agricultural technology which are made when assessing tolerable soil losses. Potential productivity losses are masked by new technologies, such as chemical fertilisers and improved crop varieties (Halcrow, Heady & Cotner 1982: 254). If such advances are to be made and put into practice by the people working the land, then higher soil losses can be 'tolerated'. The degradation of many soils such as those in East Anglia, England, to the extent that they are scarcely more than a physical retention medium for chemical fertiliser and moisture (Kirkby 1980), does not have the same social and economic impact as degradation of soils where the land users do not have, and may be predicted not to have in the future, the resources to make good the degradation by the application of massive doses of fertiliser (see also Heathcote 1980, Rennie 1982). However, even in advanced capitalist countries, the economic effect of degradation and erosion may not be negligible. Pimental et al. (1976) have estimated that erosion has reduced the production potential of American cropland by 10–15 per cent, and that an estimated 5 gallon equivalent of fuel per acre is being used to offset past soil losses, which amounts to about 4 per cent of total oil imports in 1970 (p.153).

In summary, four crucial questions to be raised in discussion of the positions of Simon, Beckerman and Ruttan are as follows:-

(a) Will future agriculture and pastoral technologies be induced which will protect fragile and/or eroded environments or compensate land users for productivity decline due to environmental degradation?

(b) If these technologies are developed, will they be accessible to current land-users of these areas?

(c) Is there any evidence now of (a) and (b) occurring in the way that Boserup and Ruttan have indicated, and that they will continue into the future?

(d) What level of soil erosion need occur before appropriate legislation is implemented and appropriate agricultural and pastoral technologies 'induced'?

So it can be seen that what was originally a technical question of the definition of tolerable rates of soil loss is a crucial, and deeply ideological one, and therefore not amenable to the cut-and-dried standards of proof on empirical grounds. However, let me put a view which though not provable, is perhaps persuasive – even if persuasion is the lowest form of proof!

The belief that agricultural technologies will be developed for marginal, ecologically fragile areas, *and* for the marginal poor farmers and pastoralists that live there is, it will be argued, a trifle heroic. The examples of induced innovation in Japan and USA responding to the different and changing relative factor prices are the best available. Induced innovation did occur for these great and developing nations because it was central to their national development and they both had particularly advantageous social and natural endowments at the time. Simply, there was money to be made in research and development (both public and private) in these strongly growing economies. Of course these nations induced appropriate agricultural innovations. The development of the Green Revolution is another case in point. This was undoubtedly a breakthrough of major significance, but it was developed with particularly strategic and world political objectives in mind: the containment of communism by increasing food production, and the reduction of US food-aid, were major impelling factors. It provided unparalleled opportunities for transnational companies manufacturing chemical fertilisers, pesticides, tractors and pump-sets to sell their wares to a vast new market. Now, the issue here is not whether this was 'good' or 'bad', the motive humanitarian or base, but that it had a tremendous impact upon the agriculture of the Third World at least in Latin America, South and South-east Asia. An excerpt from Lester Brown explains these opportunities which the Green Revolution offered multinational corporations:

> Given the pre-eminence of the multi-national corporation in international development, it is surprising that it is accorded so little attention in development literature.... Although there are occasional highly publicised instances of expropriation... foreign private investment is growing rapidly. According to the *Survey of Current Business*, American private investment abroad totalled $17 billion in 1930 and had reached only $19 billion in 1950, but then began to climb rapidly to $50 billion in 1960 and $87 billion in 1966.... Thus far only a small fraction of overseas investment of multi-national corporations has been allotted to agribusiness in the poor countries, but the amount is increasing. Sales of farm inputs and opportunities for new investment in food processing and related activities are increasing in the poor countries in close relation to the acreage planted to the high-yielding varieties. (Brown 1969: 64, 65)

The access of farmers of lesser developed countries to this package of new inputs, as is now abundantly clear, was limited. It was the larger farmers who gained in practice, although there is still disagreement over why (whether seeds and fertilisers are scale neutral, but other inputs are not, or whether high yielding varieties (HYV's) are less robust and larger farmers have the necessary economic and political power to control growing environments better). However, the creation of a new agricultural bourgeoisie in almost one of every country of non-communist lesser developed countries, as a result of the Green Revolution, has assured a self-generating demand or induced innovations of a particular type, which encourages both larger surpluses and mechanisation. This growth of biological and mechanical agricultural technology may also have benefited small farmers and the landless, but only incidentally. In other words it was not the demand signals of the small farmer and landless that generated the research development and diffusion of the Green Revolution technology but the interests of large multinational companies, and strategic considerations of the United States of America.

The crucial area of debate centres upon whether the signals from ecologically fragile and marginal areas will reach institutions and governments so as to induce appropriate innovations for those areas. While relative factor scarcities may have provided the required signals where there is a lucrative profit to be made, it is not the case for fragile areas and the small farmers who live there. Very frequently the signals from large farmers and highly productive areas lead to competition for resources with those from small farmers and relatively unproductive areas. It would be wrong, however, to dismiss completely advances in productivity-enhancing research which can be applied to steep-sloped and marginal areas. For example, in Caquezo and Rio Negro in Colombia corn and potato yields have improved 200 and 90 per cent respectively on hillside farms which were able to adopt the improved technology (Sepuldeva 1980), although this example was only on a limited scale. In Jamaica the use of hillside ditches and contour mounds produced sustainable yields of 38 t/ha/p.a. and reduced soil loss to 17 t/ha/p.a. while traditional cultivation methods only yielded 26 t/ha/p.a., with a rate of soil loss of 134 t/ha/p.a. (Sheng & Michaelson 1973).

Ruttan (1982) also distinguishes from productivity-enhancing research another type called 'maintenance research' defined as 'the research needed to prevent yield decline as a result of the evolution of pests and pathogens, the decline in soil fertility and structure, and other factors'. This quotation is one of the very few places in his work where he actually mentions soil erosion. There are a number of reasons why maintenance research is less attractive than productivity research in general, and particularly so for marginal areas. Maintenance research tends to be undramatic, unglamorous, and only missed when it is absent. It also may bring much lower returns, both economic, political and (in a bureaucratic framework) promotional, than spectacular breakthroughs in irrigated rice or wheat, for example. In the case of ecologically marginal farms, inhabited by politically marginal farmers, maintenance research is doubly unattractive. Some of these possibilities are hinted at by Ruttan in an earlier paper (1981). However, the title of this paper restricts reference to developed countries only, but even here the normative suggestions about distortions to prices of factors and outputs (leading to inappropriate signals to research institutions) seem to amount to little more than wishful thinking. 'Political dialogue' is required and a dialectical interaction at the economies' political level between research scientists, research planners, research clientele, and the legislative process is required and hoped for. His article outlines the failure of reseach programmes in Argentina, Peru and Colombia during the last fifty years (p.27) but does not explain it or help us to find a way round the problem.

By far the greatest part of maintenance research developed by institutions of state for marginal areas has been soil conservation techniques (outlined in Ch. 5), which, as this book shows, have had to be imposed by colonial powers to be 'induced' at all. However, certain aspects of maintenance research for eroded areas remain characterically neglected. For example, Spears (1982) states that little research has been done on the cost of rehabilitation of severely eroded hillsides. It is at present very expensive, in the order of

$US500–$US1000 per ha, and low cost methods have not been developed. Other examples appear elsewhere in this chapter. Of course farmers and pastoralists themselves have developed their own and usually effective methods of soil and water conservation which they had evolved themselves and suited their own social relations of production and ecological conditions, but which had been disrupted by the colonial experience (see Ch. 4 for examples). Referring principally to those techniques of soil conservation developed by research stations and government institutions, many studies of the economics of soil conservation which focus on the private economic incentives for soil conservation, show that, although *total* benefits from a soil conservation scheme such as terracing may be more than the total cost, individual farmers usually lose income from these practices (Harshbarger & Swanson 1964, Holtman & Connor 1974). Ervin and Washburn (1981) have shown that even with very favourable combinations of a low discount rate, a long planning period and a high level of individual cost sharing, private incentives are very seldom sufficient for the individual farmer. In a difficult production situation, with high perceived discount rates and competition between small producers for many scarce resources, it is obviously an unanswered question as to *who* will pay for conservation works, or how cost-sharing or insurance will be arranged on a sufficiently large scale.

Productivity-enhancing research and development in many lesser developed countries have often been harmful, where they *have* been accessible to farmers of those areas. The introduction of monoculture maize which replaced nutritionally and soil conserving mixed-cropping strategies in Africa is one example. Belshaw (1979) illustrates the contradiction between research station agronomic techniques which favour planting crops in pure stands, and inter-cropping techniques which are by far the most satisfactory from the point of view of the reduction of soil erosion. Pure stands allow accurate control of the plant population and mechanisation of weeding and harvesting (an implicit big farmer bias is evident here). They allow effective and profitable fertilizer and pest/disease control treatments which can be calculated more easily. They are congruent with the principal objective of increasing output of one crop (often for export and consumption in urban areas) rather than one to enhance the farm system as a whole. They stand for modern, improved, and mechanised farming as opposed to indigenous and backward farming, and so on. While there are many other advantages of inter-cropping than soil conservation, this type of crop planting technique is neglected and has been systematically eliminated in many lesser developed countries (see Beets 1982). In the same publication as Belshaw (1979), there are some other interesting examples of the deleterious effects of exogenous technologies upon indigenous ones in lesser developed countries many of them with strong implications for soil erosion, particularly Swift (1979) and Richards (1979). Even within one crop, varieties are often introduced in an attempt to maximise yields, but succeed only at the cost of decreasing soil protection by foliage at times of heaviest rainfall (David Gibbon, personal communication).

The emphasis of almost all research and development carried out at research stations (and economic studies for national agricultural strategies as well) is upon particular commodities isolated from their social, economic and environmental context. Farming systems research which studies the complex of decision-making by farmers as a whole system implies that the researcher spends time with the farmers rather than in the research station and adapts research needs on the spot – in the farmers' fields – to ongoing work by the farmers. The problems that arise from the development of a new crop on a research station and the attempts to recreate it in economically and environmentally diverse farms outside can in this way be avoided (CIMMYT 1980, Biggs, 1981). Environmental protection can easily be built into adaptive research, and becomes just another important but related consideration in induced innovation. But it is just this sort of research which is hardly being practised at all (with a few exceptions: Conyers 1971; Okigbo 1981; and a bibliography on farm systems by Gilbert, Norman & Winch 1980). However, adaptive research is rather demanding upon scarce skilled manpower, and it also requires researchers to live for long periods outside the comfort and orderliness of the research station itself. For this reason alone, it is not very popular with private state-sector research institutions.

Instead it is single cash-crops which attract most research, developed outside the context of the production situation in which small farmers in steep-sloped and ecologically fragile areas find themselves. Thus the introduction of many cash crops, particularly peanuts, coffee, cotton and maize into ecologically fragile areas has had disastrous effects (and some of these are analysed in Chapters 8 and 9). The existing biases of research and development are well known and include big as against small farmers bias; cash crop as against subsistence crop bias; maximum average yield as against maximum yields in bad years; irrigable as against dry land crops; and higher yields per unit as against higher yields per worker. Innovation is induced very healthily along the lines of *these* biases as Ruttan suggests, but not along those enabling small farmers and pastoralists to maintain *or* transform their livelihood in areas of highest environmental vulnerability.

Three pieces of evidence noted by Posner and MacPherson (1981 fn.) support these arguments. First, The World Bank estimates that between 90 and 95 per cent of public investment in agriculture since World War II in Mexico and Peru has been in irrigated agriculture, and practically none in watershed management. Second, the major impetus for colonisation schemes in Colombia, Ecuador and Peru was *not* to solve the problems of the pressing needs of the hill areas at all, but to increase aggregate agricultural production and reduce rural-urban migration without causing political unheaval as a result of land reform. Third, a flat-land bias in research stations in tropical America is very marked. So far research has focused almost exclusively upon export crops such as sugar, bananas, cotton and beef and products consumed in urban areas to the neglect of inter-cropped maize, beans, potatoes and barley. Also most land classification systems put a ceiling of 10-15 per cent slope for annual cropping, thereby diverting researchers' attention away from the problems confronting hill farmers.

The next part of the discussion moves away from research institutions to farmers and pastoralists themselves, and concerns the accessibility of new agricultural technology to these people. Yield-enhancing innovations are usually the only ones available through the network of state institutions. However, these can only be taken up by larger farmers with access to credit, with a lower vulnerability to the risk that the innovations imply, and with the economic and local political power to assemble all the necessary physical inputs together in time. If these new inputs had been in a form that small farmers in marginal areas could practically use, it could be hypothesised that the resulting extra incomes would encourage soil-conserving investments and would take the pressure off farmers to over-exploit the environment. However, as Chapter 7 explains, yield enhancing innovations, particularly concerning cash crops, both tend to exacerbate soil erosion and accelerate differentiation of the peasantry (see Ladejinsky 1969, Griffin 1974; Pearse 1980). Other examples of applications of high energy technology to agriculture and stock-rearing, such as deep boreholes in semi-arid areas are discussed by Cliffe and Moorsom (1979), and Kassapu (1979), where it becomes even more difficult for those with dwindling resources to avail themselves of soil conserving and yield enhancing new technologies. An atomised peasantry of small farmers is one of the least attractive and the most difficult target populations for state-sponsored innovations (even if they are appropriate). First, small farmers are many – while bigger farmers tend to be few. Second, they do not seize opportunities for credit and new inputs, while large farmers are frequently able to look after themselves even without the aid of extension agents. Small farmers too have limited room for manoeuvre to adopt new technologies. The rural sociological literature of the 1960s of the diffusionist school has demonstrated to the point of overkill the problem of the laggard and the small farmer (Rogers & Shoemaker 1971). Environmental decline further limits small farmers' and pastoralists' ability to accept externally generated innovations. This vicious circle is more precisely specified in Chapter 7.

Internal or autochthonous innovations, the sort which Boserup (1981) discusses under conditions of population pressure, do of course occur, and her book gives many examples. However, it is striking that discussions of the state, of relations of production, and patterns of surplus extraction are almost completely absent. Without these crucial areas of explanation, her theory of how innovations occur or do not occur seems remarkably fragile. There are a number of processes which reproduce themselves through time and which keep poor farmers in perpetual poverty and force them to use their natural environment in non-sustainable ways. These processes, excepting that of population growth, are largely unexplored. While there are local adaptations of agricultural and pastoral practice to conserve soil, these processes ensure that they are usually not far-reaching, nor fast enough. The necessity for speed in adaptation of agricultural technology exists because of these self-reproducing and self-reinforcing processes – fairly rapid population growth; political and economic neglect of marginal environments except where they can be exploited for the production of commodities for the national and international market; and a stagnant production per unit worker and area for the majority of the population. The last symptom may seem tautological, but in effect it is self-reinforcing. Once levels of savings and the ability to risk innovation decline (sometimes exacerbated by malnutrition and food shortages at critical points in the agricultural calendar) the necessity to 'do something about it' becomes all the more urgent and difficult. Adaptations do occur, and the ability of peasants and pastoralists to adapt their systems of resource management are often impressive (e.g. Wilken 1974, Brokensha et al. 1980; Denevan 1980; Beck 1981). To give a brief example, in Nepal population pressure has in the last decade or so brought about many intensifications of cropping patterns such as the introduction of wheat as a winter crop. More rapidly maturing early paddy allows an extra crop of maize to be harvested in irrigable fields; double-cropping of paddy on irrigable fields is now commonplace; and inter-cropping of squashes and beans in maize fields is being introduced. But the pace of population pressure and the failure of fuel supplies (current demand is 546 kg per *caput* of firewood per year, and the maximum sustainable yield at present is 80 kg), have reduced the fertility of the soil, the supply of perennial irrigation and drinking water, and increased the rate of sheet and gully erosion (Banister & Thapa 1981: 87–94).

Whether it is a matter of a lack of knowledge on the part of indigenous cultivators and pastoralists that leads to excessive and untimely tillage, the use of improper implements, the destruction of crop residues by burning and so on, is a moot point. While FAO earlier reports (1960, 1965, 1966) and others (cited in Ch. 4.2) put these practices down to ignorance, there are many examples where knowledge is not the scarce factor, but the resources to put this knowledge into practice. My own experience in Morocco, Zambia, India and Nepal is that there is an enormous variability in people's perceptions of environmental decline. Some see it and have sound explanations, like one Magar lady in rural west central Nepal who gave a ten-minute lecture to me on the problems of the transference of fertility from forest to arable land, and changing uses of crop residues with increasing population pressure, that would have stood up well in a graduate seminar. Others in the same country seem to have little notion of what is happening (e.g. Bajracharya (1981) from eastern Nepal). Certainly no generalisation can be made except that the successful diffusion of the knowledge of soil conserving practices is often *one* essential change that is required – even although the knowledge may be possessed by some cultivators already. Usually, as this book will argue, successful agricultural extension is a necessary but far from sufficient condition for implanting sound conservation practices (see Thrupp (1981) for a case study of the (jaded) peasant view of conservation in Costa Rica).

Solutions to population pressure, differentiation, and declining productivity through soil exhaustion and erosion, which lie outside the agricultural sector altogether, do not offer the same promise as they did in Japan, the USA or the United Kingdom during their industrial revolutions. Outmigration to cities in search of menial jobs or part-time work in nearby plantations provide remittances and reduce food demand, but these activities also reduce labour availability and have been responsible for the collapse of old-established terracing systems in the Middle and near East, particularly Yemen and also in the Andean part of Peru and Colombia. In some ways these experiences both alleviate the problem as well as reproduce it, since it enables some outside income to prop up a failing agricultural economy postponing either abandonment of the area and/or more radical transformations.

It is argued in this book that in many areas of the world where environmental fragility is an outstanding characteristic, there is a failure to adapt to a variety of new and related pressures, particularly population pressure and increased state intervention which is often extractive in nature, and also that such technically state-sponsored innovations that there are, tend to be inappropriate or inaccessible. Thus there are twin reasons for this failure to adapt in these areas. The first is from outside concerning the State and the second from inside, in that the inhabitants cannot adapt their own technologies fast or radically enough to maintain an increase in their incomes, nor can they find a solution elsewhere by migration and a secure source of income outside the region altogether.

Consider now Simon's remark (1981: 84) that:

> even such persons who worry about the 'loss' of land as Erik Eckholm, acknowledge that it is in our power to have more land if there is a will to work for it. Today the human species has the knowledge of past mistakes, and the analytical and technical skills to halt destructive trends and to provide an adequate diet for all using lands well suited for agriculture.

To whom is the word 'our' addressed? The argument made in this book is that it is not to all land-users. Neither is it a question of 'working for it', nor, to put it the other way, that people do *not* have more land because they are *not* prepared to work for it. They are excluded from using more land, and they often work onerous hours on the land which they do own for a pittance. Lastly to talk of the human 'species' fails to differentiate between people whose circumstances are very different. The approach in the opening quotation is the same as that taken in this chapter 'taken as a whole ...'. Here we would never maintain the whole is suffering environmental decline, but that parts of it are, and this matters to the numerous people living there. Aggregates tend to appeal to (comfortable) analysts.

There is a further sense in which a lack of access by land-users renders the aggregate view very questionable. So far the political and economic aspect has been emphasised, but there is also a spatial one. There are barriers to a free movement of people from areas of high to low population pressure. Some of these can be expressed as a distance-decay function and indeed costs of migration may largely be a matter of distance as in the case of outmigration from island economies such as are found in the Caribbean or Oceania; others are political and virtually insurmountable, such as national frontiers. The case of India and Nepal is a case in point. Population density on the Nepal side of the *terai* is four times that on the Indian (Gaige 1975), but the physiography and soils are similar on either side of the border. Another example is highlighted by some extraordinary satellite pictures (published in Glantz 1977: 6, 7) showing the difference in vegetative cover and population pressure (both human and animal) between the Republic of South Africa and Lesotho. Moreover, the creation of political boundaries in the Sahelian countries during colonial and post-colonial times has actually contributed to the disruption of a carefully regulated system of pastoral movements and pastoral-agricultural symbiosis, leading to environmental deterioration and desertification (Franke & Chasin 1980: Ch.3 & p.98).

Policy-makers and commentators in many countries of the world have often assumed that small farmers and peasants in mountainous and/or environmentally degraded areas will abandon their hillside plots if given other opportunities elsewhere, so that the state by various means can ease population pressure on steep slopes by encouraging or at least not inhibiting colonisation of new areas. In fact hillside farmers tend to be very attached to their farms. In many cases they have, through time, been attempting to subsist on farms that have become progressively smaller without becoming more productive. However, they have consistently opted to eke out a living by temporary migration and wage labour rather than leaving in such numbers so as to exceed natural population growth, and thereby ease the pressure. Colonisation schemes in many tropical American countries for example, have also proved to be extremely expensive (e.g. $US6000 per family in Bolivia, Curtis 1979), as is also the case of Indonesian policy of migration from Java to the outer Islands (see Booth & McCawley 1981 for detailed accounts of the failure of these schemes). The traumatic effects of resettlement, whether as a result of dam projects or a state-sponsored attempt to relieve population pressure, often reduce the ability of the settlers to adapt satisfactorily to their new environment.

In other cases there *is* land to spare and at the same time extreme population pressure plus unequal landholdings, such as in Brazil, where there is a constant stream of immigrants to the marginal lands of the North-East and to Amazonia. These migrants are either middle or large 'farmers' who can obtain credit for farming on very easy terms from a variety of government institutions (and in the hyper-inflationary economy of Brazil, easy credit amounts to a free gift of capital). Then the immigrant purchases a large tract of land, and farms it without any expenditure on soil conservation, often without chemical or organic fertiliser, until yields have declined and degradation or erosion has set in. He then abandons his farm, raises new loans and moves on. Other immigrants are less fortunate and have very little capital. For them there is frequently a cycle of desperate and under-capitalised farming followed by failure through drought or poverty-induced disease and malnutrition, with another enforced move to take more land into cultivation, while the land under previous cultivation erodes, or reverts to unproductive grazing land (Gilbert 1974, Deutsch 1977: 343 f.; Johnson 1978; Grainger 1980: 16). The same process may also be happening across the border in the North East Oriente of Ecuador (Hiraoka and Yamamoto 1980). However, these examples of Andean countries Brazil and Indonesia are all ones where there is land still available. Colonisation (either planned or unplanned) and removal of population pressure are simply not possible at all in small sea- or land-locked countries (e.g. in the countries of the Sahel, Lesotho, Swaziland, Rwanda, Burundi, or the islands of Oceania). In any case large-scale planned colonisation schemes have a very low success rate.

The last point about the relationship between farmers themselves and their natural environment concerns the production and preparation of food (including the cooking of it). Here again it is a matter of ideological assumptions and the unit of study. Trends in food production by continent, for example, show an increase except for Africa (FAO annual publications entitled *The State of Food and Agriculture*). Here the level of aggregate production, even at the national level in many cases is not the problem, and I agree completely with Simon and Beckerman's repudiation of neo-Malthusianism and the Club of Rome's doom-dominated world model. It is widely recognised that it is not production of food but the low level of demand for food that is the problem for many and for varied reasons related to the forces and relations of production in which people are involved. A combination of vulnerable classes or groups living in vulnerable and ecologically marginal areas in many cases causes a failure to create sufficient income for food to feed themselves. Again, it is, as Sen puts it, 'the tradition of thinking in terms of what *exists* rather than in terms of who can command what' (Sen 1981: 8, quoted in Redclift 1982: 1) which ignores the vital question of access to food production and the fuel to cook it. As Agarawal (1980: 150) in terms of wood fuel puts it, in the same way as Sen, it is not a simple technical problem but:

> It is a problem of poverty and political economy. It is an issue both of absolute shortages of the wood ... and the distributions of available wood supplies between different uses and users ... its causes are seen to be closely linked to the distribution of economic and political power between classes and social groups which determines who gets how much of a scarce resource.

Many groups of people suffer from stagnant or falling rates of both food and fuel production alongside others who have very significantly increased theirs. This is also reflected in an aggregate geographical sense where whole areas suffer from this problem, while others have markedly increased their share. Here a hypothesis on *regional* differentiation, sometimes coinciding with social differentiation is attractive, although it is very difficult indeed to provide data to support it. Areal data will not suffice by themselves, since it is quite possible for aggregate areal food production to increase in such a way that increasing numbers of people have difficulty feeding themselves. Also 'areas' may refer to a few kms^2 or to whole countries, so that areas or pockets of declining food production may coexist with, and have no social access to, other areas of increased food production. In these circumstances soil erosion is potentially of utmost importance to the livelihood of these vulnerable people, even if it is not happening at present. As we shall suggest later, the relationships in which the inhabitants are enmeshed often encourage soil degradation and erosion in fragile environments – which has the effect of a vicious circle and makes it even harder for transitional and progressive technical (and political) changes to be made. In the words of Gallopin and Berrera (1979); 'They (the poor) may be forced to destroy their own environment in attempts to delay their own destruction.'

A similar view is put by Allen (1980: 29). Eckholm (1982) identifies a global underclass, produced and reproduced by social relations of inequality, which is an indictment in itself, but which also causes environmental problems across a wide range (not only soil erosion, but also urban pollution and disease). The ways by which this class comes about and the implications and feedback effects to and from the natural environment are discussed in Chapters 6, 7 and 8.

Three brief case studies illustrate the related issues of population growth (without significant outmigration), stagnant production, poverty of most of the people of the area concerned, and environmental decline.

The first, is Rwanda where 'given conditions of rapid population growth and limited natural resources, Rwanda faces a number of serious environmental problems' (USAID; 1981 p.V). Deforestation, overgrazing and shortened fallow periods threaten the fragility of Rwanda's soils. One illustration of the extent of the damage is that in the Eastern Plains where domesticated animal off-take is 900-1300 kg/km^2, that of the wild ungulate biomass in unspoilt areas of a national park is in the order of 15,000 – 28,000 kg/km^2. Population pressure on land has led to increasing competition of less sensitive rotations which maximise the production of food crops with beneficial rotations which include nitrogen-fixing crops and forage crops. An agricultural revolution of major proportions is required involving widespread (and costly) irrigation, reafforestation, land management practices including agricultural diversification and erosion works – otherwise an annual population growth rate of 3 per cent and an economic infrastructure which cannot make productive use of the growing population will precipitate a major crisis – and also accelerate soil degradation. Fig. 2.1 illustrates these points. If farms remain the same size as now (1.1 ha), all farmland will be saturated by 1995 at a population of 5.7m., marked 4 on the diagram. But existing mean farm size is too small for a vigorous, self-sustaining 'take-off' (which was estimated to be possible at a minimum farm size of 1.8 ha). Under these conditions, saturation has already occurred (in 1977) at a population of 3.8 m., marked 5 on Fig 2.1. Increasing failure to grow enough food on the part of farmers leads to a decline in productivity through a lack of energy and malnutrition particularly at times of peak labour demand, and an attempt to substitute less soil-conserving crop rotations.

The second case study is Swaziland. Tables 2.1 and 2.2 from Roder (1977) tell a sombre story.

Fig. 2.1   Population pressure in Rwanda (from USAID 1981)

① Estimated total population of Rwanda (growth rate c. 3% p.a.)

② Population that could live from farming, given a linear projection of population growth in relation to a farm size of 110 ares (as at present)

③ Population which could live from farming under 'take-off' conditions in relation to a farm size of 180 ares

④ Population which could live on farms of 110 ares

⑤ Population which could live on farms of 180 ares

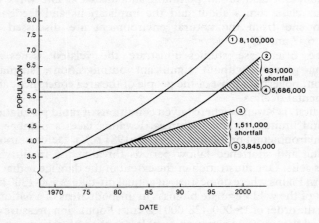

Source: from MAB 1981, p. 85

Table 2.1   Current and optimum stocking rates in Swaziland (Roder 1977)

Region	Current rate (ha/animal unit)	Optimum (ha/animal unit)
High veld	1·94	2·65
Middle veld	1·33	2·65
Low veld	2·10	4·00

Table 2.2 Grazing intensity and life expectancy of soil, Swaziland (Roder 1977)

Estimated grazing intensity (ha/animal unit)	Area affected in Swaziland (ha)	Life expectancy of soil in year
0.40 - 0.75	16,790	less than 5
0.75 - 1.00	55,220	5 - 10
1.00 - 1.25	65,200	10 - 15
1.25 - 1.50	91,600	15 - 25
1.50 - 2.00	79,355	25 - 50
2.00 - 2.50	256,190	50 - 100
Over 2.50	690,165	100

Efforts to reduce stocking densities have met with little success due to the significance of cattle as a source of wealth and prestige rather than a source of income (IBRD 1977a; Doran, Low & Kemp 1979), so that herd size has actually increased between 1968–77 in spite of government efforts to reduce it (this being discussed further in Ch. 7).

The last example is situated in tropical Latin America and is taken from Posner and MacPherson (1981, 1982). The proportion of the national area of thirteen tropical American countries described as steep sloped is very high – 80 per cent for Honduras, Panama, Haiti and the Dominican Republic and not less than 50 per cent for eleven of the thirteen. Also these steep-sloped areas are very important contributors to the national total production of foodstuffs as well as fibres and export crops (particularly coffee). Many of these areas also suffer from extremely serious soil erosion, although reports and impressions seem to differ. For example, Eckholm (1976b) estimated that 77 per cent of El Salvador had severe erosion, and Low (1967) estimated that 87 per cent of Peru was losing more than 10 t./ha/p.a. of topsoil. Baldwin (1954) states that 42 per cent of Mexico had 'accelerated erosion' and so on, although as section 1 of this chapter suggests, it is difficult to know how to gauge the significance of these figures except in a general and qualitative way, which indeed may be sufficient. Posner and MacPherson (1981) show that staple crop yields have tended to decline (Blasco 1979) or at best remain

Atlas data?

stagnant (CEPAL 1976). Population growth in the steep-sloped areas of these countries is expected to increase by the year 2000, by between 40 and 59 per cent in many areas (e.g. Ecuador, El Salvador, Guatemala and Honduras). The resulting situation is therefore very serious – both in a national sense, since these areas contribute significantly to food and export crops, and also in a regional sense since erosion, population pressure and poverty will probably increase, given existing trends and national priorities in research and investment.

These examples do not 'prove' a case, but they help, and there are many well-documented cases in which a variety of factors combine to produce evidence that erosion is important to land-users and to the national economy, and that neither research and development, nor local adaptations in agricultural technology will provide answers to these problems. Lest the reader may point out that the examples so far are predominantly drawn from small countries and affecting only small populations, others, such as areas of China and India are discussed elsewhere in this book.

3. Social elements in soil erosion

Soil degradation and erosion is caused by the interaction between land use, the natural characteristics of that land and its vegetation, and the erosive forces of water and wind (Stewart 1970). The focus of this book is upon the social element, but not to the exclusion of the physical parameters. Clearly any case of soil erosion must be the result of this interaction and any *ceteris paribus* assumption about these physical parameters would be absurd. The central question asked is why certain land-uses take place. Why are no soil conserving practices a part of the on-going development and change of land-use practices, and why are adaptations to conserve the soil not sometimes made? The answer to these questions lies in the political-economic context in which land-users find themselves. The analysis must start therefore in areas which initially may seem remote from the physical processes which directly cause degradation and erosion.

One of the main problems in the discussion of the political economy of soil erosion is that it is difficult for author and reader alike to span the great variety and complexity of circumstances (both physical and social) under which soil erosion occurs. For example, much of the foregoing discussion was making the point that soil erosion *was* important to many small producers in steep-sloped and/or semi-arid areas and it may have seemed that these people were the only ones to cause and suffer from the effects of soil erosion; also that there were few or no cases where successful adaptations and advances in agricultural technology had occurred. Clearly this cannot be the case. Therefore generalisations made in the course of the book usually must be qualified heavily by such convenient disclaimers as 'sometimes', or 'under certain circumstances'. So that some of this variety can be displayed in a simple fashion at the outset, a listing of different political economic contexts under which soil erosion occurs is given below. Lists are not meant as explanations, although there is sometimes an implicit taxonomy and even theory in a list, and the theoretical starting point of this book emphatically does not lie in this list. A cursory examination shows that the items on the list are not mutually exclusive, and thus the criteria used for compilation are mixed and also not mutually exclusive. The main purpose of the list is to illustrate complexity and variety of political-economic and physical circumstances of soil erosion. The principal criterion used in the identification of different social elements in soil erosion is the relations of production under which land is used (and these are more carefully specified in Chapters 7 and 8).

(a) Soil erosion and degradation can occur in peasant and pastoral groups. Here family labour is employed, usually under the direction of a senior (and usually male) member of the household, and production is primarily for use rather than for sale, although the latter may be economically essential. Some of these groups may be relatively undifferentiated, although considerable inequalities in access to land, pasture and family labour power usually exist. Other important variables which qualify the impact of these relations of production upon the lives of these groups include the availability of land, whether there is population pressure under the existing production technology, and the degree of interference by the state (by way of taxes, conscription, price fixing, rural development projects and so on).

(b) There are also peasant and pastoral groups that are characterised by marked differentiation, where a considerable proportion of the means of production is owned by another class altogether. They can be *hacienda* owners where feudal relations with workers on the *hacienda* may still exist, or capitalist entrepreneurs (either owners or managers for very large corporations and transnational companies). The social roots of soil erosion under these circumstances are more complex. They derive from both the operation of large units (often highly mechanised and characterised by monoculture and high energy demands sometimes supplied from local sources); and the small peasantry itself which has to work for part of its income on larger enterprises. The latter often are pushed off the most suitable land for cultivation and are forced to make a living cultivating steep slopes or pasturing animals on a reduced amount of land.

(c) Soil degradation and erosion also occur in centrally planned economies, where decisions are made to use land in a non-sustainable manner. This arises from a number of causes. A Bolshevik (and, it might be argued, a Marxist) ideology in which man's mastery over nature is essential in the rapid development of the forces of production is central here. Also there has tended to be in earlier Soviet writing a sort of radical cornucopianism which has provided the ideological context of some technically disastrous decisions on the use of land. Consider this stirring quotation and its possible ecological implications:

> We must discover and conquer the country in which we live. It is a tremendous country [referring to the USSR] but not yet entirely ours. Our steppe will truly become ours only when we come with columns of tractors and ploughs to break the thousand-year old virgin soil. On a far-flung front we must wage war. We must burrow the earth, break rocks, dig mines, construct houses. We must take from the earth... (Burke 1956, quoting from *The Great Plan* (1929)

Pryde (1972) draws attention too, to the assumption of Marxian economics (but not perhaps all Marxian economics) that only labour produces value, and therefore that land, water and all associated resources are considered 'free' inputs in production. This has led to some very unfavourable repercussions of natural resource use, for example the extension of wheat production into Kazakhstan under President Khruschev. However, ideology apart, there is considerable evidence of a degree of senior bureaucratic bungling and the suppression of scientific reports which would would have been embarrassing to the senior members of the bureaucracy if Komarov (1981) as the chief informant is to be believed, along with other writers such as Cherenvisinov (1964), Goldman (1972) or Kirby (1972). It is difficult to sort out blatant public relations exercises from virulent anti-Soviet propaganda. Although every ideology forms its own criteria and language of evaluation, it is doubly difficult to be discriminating and to carry through a particular form of objectivity when reading accounts of conservation in socialist or 'socialist' countries. (The inverted commas here indicate that there is considerable debate about which countries can be called socialist.) A reading of Pryde (1972) or Tregubov (1981) simply is not factually consistent with that of Komarov (1981). Here one suspects that there is not so much a clash of ideology as one of gross distortion.

(d) Erosion also occurs in advanced capitalist countries. Here farms are privately owned (either by owner-operators or larger corporations), wage labour is employed, and the product is grown for the market. Usually agricultural technology is closely linked to that of industrial technology and is constantly innovating as private farms strive to remain competitive. The direction of the technology is one of control over nature, yield maximisation, and at the same time reliability of quality, control over yields and harvesting dates; as a whole it responds quickly to the needs of the market. Land use itself also quickly adapts to price signals, and the economic margin of cultivation can therefore change very rapidly. However, if farms remain profitable, yield-increasing technologies tend to mask the effect of soil degradation and erosion and make up for declines in fertility that would have occurred if land had been cultivated with a constant level of technology.

There are a large number of other contingencies which crosscut these four political-economic contexts and modify them. Indeed, the existence of soil degradation and erosion is contingent on an enormous variety of human and physiographic variables, and it is probably impossible and counter-productive to attempt a single theory of soil erosion. Instead this book attempts to theorise substructures in a theory of soil erosion. A number of contingencies discussed below are illustrative and not exhaustive.

Rural population densities are of prime importance in modifying the implications of the different relations of production under which land is used. Frontier areas with low population densities attract particular kinds of exploitation of natural resources. Large logging operations for the export of timber, or to supply local plywood pulp and construction industries, frequently do not bear the costs of ensuring sustainable yields and often 'mine' the natural biotic resources of areas and then upon their exhaustion, move on (see Ch. 8). The phenomenon of the hollow frontier has been common where settlers cultivate land, which is virtually free to anyone with the capital and/or labour to exploit it, and move on when serious erosion sets in, leaving behind eroded soils. In the history of the settlement of the United States the first white settlers in north America arrived in Virginia in 1607, and by 1685 there was a serious increase in flooding caused by forest clearance. With rapid depletion of fertility in the Tidewater land many settlers moved onto the Piedmont where they mostly repeated their mistakes. The expansion of settlement soon became based on a hollow frontier as settlement moved west leaving a trail of erosion and siltation behind until by 1939, Charles Kellogg felt that 75 million acres (28 million hectares) of this [(450-500 million acres/180–200 million hectares of eroded land)] were too worn out to return a living wage under any system of farm practices. (Kellogg 1941). Similar settlement histories have occurred in the Gangetic plain and Doabs of northern India (Schlich 1889: 187-238) and Brazil up to the present day.

High population densities on the other hand may modify the

way in which the land is used under given social relations of production, and indeed modify the relations of production themselves. Where there is no further land for settlement and for people to set themselves up as independent farmers or pastoralists, there is an added reason for a land-controlling class to emerge, firstly because land hunger tends to differentiate a peasantry and secondly there is no alternative for those without enough land but to work for others. However, high population densities are certainly not a prerequisite for inequalities and a class structure to arise.

Although population pressure is the most common attribute of these peasantries and pastoralists, there are many cases where declines in population through outmigration or apocalyptic disasters (e.g. Latin America during and after the Spanish Conquest, Boserup 1965: 62–63) have led to a coarsening of agricultural technology, a labour-saving cropping pattern, and an allocation of labour which can encourage soil erosion (e.g. Bunyard 1980 for Palestine). However, in these cases too, it is the economic relations which farmers have with other more developed economies that cause these problems.

The nature of the state in which land-use occurs can also modify the broad classification outlined. The type and degree of infrastructure such as roads and storage facilities, institutions such as banks, agricultural extension development projects, fiscal and monetary and pricing policy, the legal structure of land tenure and of environmental protection and conservation; the state's degree of intervention in international imports and exports – all these are part of an explanation of soil erosion.

In the analysis of all specific cases of agricultural technology, crucial aspects which may lead to soil erosion, cannot be 'read off' in a vulgar materialist sense from the relations of production under which they are applied. First of all, there may be contradictions between the social relations of production and the agricultural technology used, therefore at a particular moment in time technology and social relations may seem out of adjustment. For example a semi-feudal landlord in a developing country today may employ a large number of retainers or family servants who will work for cash rent, labour rent or a proportion of the harvest on his land. This arrangement is an important part of the social relations of production. However that same landlord may be introducing tube wells, tractors and high-yielding varieties of seeds, and therefore wish to get rid of the large number of inefficient workers to whom he may be partly responsible for housing, clothing, looking after their aged or sick, and also to whom a sizeable proportion of the additional benefit from these improved inputs would accrue. This is a contradiction between relations of production and technology. To take another example, peasants often form reciprocal labour groups to cope with periods of particularly high labour demand (e.g. paddy transplanting, or preparing plots for burning in the forest for shifting cultivation). However, many of the more advantaged peasants (either in terms of household labour power, control over land, or number of bullocks for ploughing) might like to grow crops for the market. A change in crop mix, perhaps also in crop planting technology and field preparation has strong implications for soil conservation, but the additional inputs required, particularly of labour, put strains on the reciprocal labour arrangements. Here again we have a contradiction and a society in transition (e.g. for India, Saith and Tankha 1972). In short there is a dialectical relationship between agricultural technology and relations of production. That potent and overworked word 'dialectical' can be explained (after Howard & King: 1975, 21) in the following way:

> '...the main idea is that social phenomena are seen as existing in relation to each other, and continually developing in and through such relations so as to form at various phases contradictory forces that generate qualitatively new formations. Thus social reality is seen always as in a state of becoming something else.'

The second reason why a simplistic relationship between agricultural technology and relations of production cannot occur is that there are frequently many important local variations in technology which can only be uncovered by detailed analysis of settlement histories and cultural development. The path which an agricultural society will move through history, particularly with regard to the development of its agricultural technology, is determined by a highly complex reality. Relations of surplus extraction, population growth and the relevant characteristics of the environment are only three sets of variables amongst many others. Further development of Boserup's thesis is required to uncover the reasons behind environmental degradation on the one hand, or successful intensification of land use on the other (for a review of some of the major issues, see Brookfield 1982).

In conclusion, there is a wide variety of social and physical contexts of soil erosion. The social relations of production under which land is used is a key and pervasive element in the explanation of soil erosion; it also goes some way to explaining the nature of the state – which intervenes and influences the use of land in all sorts of ways. However, there are many cross cutting contingencies which have to be analysed fully in an explanation of any concrete instance.

Chapter 3

A review of techniques and policies

1. A distinction between techniques and policies

At the outset there is an important distinction to be made between techniques on the one hand and soil conservation projects, programmes or policies on the other. A technique refers to the mechanical or agronomic method applied, and in any one conservation project there may a considerable variety. The choice of technique is a vital one in a conservation project, programme or national policy for two related reasons. First, it will determine the technical success of the project in a *ceteris paribus* sense – that is to say, that provided there are no social constraints upon the successful application of the technique, it will fulfil the objectives of the project or policy (typically in terms of reduced erosion, increased yields of crops, pasture or forest). Second, the choice will impinge upon the social, economic and political life of the direct land-users involved, as well as village leaders and local and national politicians, and involve administrators, extension agents and other government personnel in particular ways.

The distinction between technique, and project or policy is clearly drawn from this evaluation of conservation in Tanganyika:

> We have already seen that most conservation schemes failed in Tanganyika. However the reasons for their failure cannot be found simply in an examination of the physical factors of the soil and environment. With some notable exceptions such as the use of terracing in the Ulugurus, most of the measures involved could have been successful in physically preventing soil erosion. However, the colonial administrators were wrong in their particular approach to the problem and it is here and in the reaction of the Africans to colonial attitudes that we can find reasons for failure. (Berry & Townshend 1973: 250)

A conservation policy has a number of components. First, a method of identification of the geographical areas in need of conservation has to be devised (e.g. land classification), as well as the identification of other associated causes of soil erosion that may lie outside these areas altogether (e.g. land tenure legislation). Second, it has to be decided who chooses these areas (the farmers, the district governor, the foreign consultant, or a committee in the Ministry of

Agriculture), and likewise, who chooses the criteria for the choice itself. Third, the choice of conservation techniques and/or other policy measures is another important component of a conservation policy. Fourth, the means of implementation is vital in terms of who does the policing of fenced-off pastures or restrictions on land-use, and how extension, new inputs and other specialist skills are to be delivered to the farmers or pastoralists themselves. Fifth, the question of who pays, who benefits, and who loses is an essential one to be addressed by any conservation policy. Lastly other related policy issues at the national level, which usually involve wider considerations than soil conservation are an intergral part of any conservation programme. Thus a conservation policy involves a wide ranging set of economic, political and social issues.

So it is the success of an implemented programme or policy, rather than the success of a conservation technique at a trial stage under research station conditions, which should be the chief objective of evaluation and critical analysis. This implies that evaluation of techniques should be a sub-set of more ambitious terms of reference. It is not surprising that for every evaluation of a conservation programme or policy, there are perhaps ten of conservation techniques. There are a number of reasons for this. First, there are the considerable methodological problems of economic evaluation of whole policies and programmes in terms of costs and benefits, particularly the latter (Amos 1982). Second, trials of conservation techniques usually originate under controlled conditions of the research station, or large plantation, where there are facilities for detailed and undisturbed instrumentation. This further encourages the evaluation of techniques in terms of physical targets rather than of other wider objectives in actual practice (i.e. in a sizeable project, or programme).

There is a considerable number of evaluations of pilot or small projects involving a small catchment and a handful of farmers (such as those undertaken routinely by FAO, IBRD, or other bilateral aid agencies). These pilot projects represent in principle a transition phase between research station and the real world. However, one suspects that the transition is illusory since the formidable constraints to translating techniques into widely-adopted practice only become fully apparent when the project is big enough potentially to have a significant impact. It then runs into problems of political opposition of local people, enforced differentiation between winners and losers in the countryside, and may expose the inadequacies of the agricultural extension staff, agricultural engineers, support staff and the like. In short, replication often becomes impossible. This general tendency in soil conservation evaluation derives from similar problems identified by critics of the existing approaches to agricultural research, namely the lack of a continuous outreach to and from research stations and farmers, and the conducting of programmes under conditions that do not exist for the farmer/pastoralist who is supposed to adopt them (Biggs 1981). Third, in many cases policies are not evaluated because they often tend to be unsuccessful. Admittedly there is a sense in which this assertion is tautologous – if they are so seldom evaluated, how can one say they are often unsuccessful? As the next section attempts to show, the case is largely made in reverse – there is very little written evidence of successful programmes and sizeable projects, and most partial evaluations and comments on conservation policies indicate failure. The point here is that techniques get evaluated and researched but programmes and policies do not, with the result that the 'social factors' which block policies tend to go unresearched too, or at best identified piecemeal, although lip-service to their importance is sometimes paid. It is an almost universal assumption that a conservation policy is a set of conservation techniques – and very little else.

The 'state of the art' of soil conservation is clearly in a sorry mess: 'The need for soil conservation was emphasised by Bennett (1955), Jacks and Whyte (1939) and many others...but so little has been achieved that most recent reports...are virtually undistinguishable in factual content and sentiment from the earlier studies.' (Morgan in Kirkby & Morgan 1980: 303).

Certainly a reading of the soil conservation literature through seventy years or so exposes the lack of substantive progress in the identification of the problems and even the technical solutions (with the possible exception of the demonstration of the importance of rainsplash in soil erosion). In the English language literature, Malcolm (1938, although apparently much of this work was written as far back as 1910), Glover (1946), Rounce (1949) and Hyams (1952) alongside those mentioned in the quotation above, are all striking, original works, marked by their intimate knowledge of local agricultural practices and the general processes of soil erosion. Recent literature has not dated these classics – an indication of a lack of its ability to transcend them.

2. The techniques themselves

This section reviews briefly the range of techniques which are available at present. Readers are directed to a number of standard texts on the subject (Hudson 1971; FAO 1977a; Morgan 1979; Webster & Wilson 1980). Although this book is not concerned directly with techniques themselves, a political-economic analysis of soil erosion and conservation must start with them. The section takes the form of a list and is meant as an introduction for social scientists who are not familiar with conservation techniques. For those who are in the professional fields closest to conservation techniques, this section will not offer them anything new.

They are of two different types, mechanical protection works and agronomic methods, although they are usually used together in any soil conservation project. Mechanical protection works are permanent structures of earth or masonry which are designed to protect soil from water erosion, and to conserve water as a resource by means of interception, diversion and deceleration of surface and sub-surface run off. These include gully control structures; waterways and storm diversion channels in or above cultivated areas (sometimes planted with grasses, sometimes floored with stones or concrete); storm dams of earthen, concrete or stone and wire construction; and terraces. The latter have a very wide variety. Indeed, contour ploughing, contour brushwood planting, and ditch-and-bund constructions could be said to be types of terrace. The oldest mechanical conservation measure is the bench terrace which consists of a series of steps cut into the slope on the contour, of which the forward edge can be lined with stone or planted with bushes, trees or grasses. They clearly require very large labour or capital inputs and are only suitable where either very high value crops can be grown, and/or in areas of high population density. They are found around the Mediterranean, particularly for vine growing; in semi-arid areas such as Yemen and Peru; and in many localities of the humid tropics where population densities are high such as the Philippines, Java, Sumatra, China, the Himalayan foothills, Rwanda Burundi, Uganda and the Cameroons (Goudie 1981: 136, 127-39, for a general review of the physical processes of soil erosion and human efforts to control it). Other types of terrace also exist, such as the ridge terrace where a ridge of earth on the lower side and a channel on the upper side provide the mechanism for increased water retention and infiltration, and reduction and deceleration of surface run off. These frequently need large amounts of earth moved, particularly the broader-based ridge terrace, and are therefore frequently built by terracers or graders. The disposal of run off from most types of terrace needs careful construction of a system of drains and channels usually designed on the basis of calculations (or informed guesses) on the probability of the intensity and duration of storms.

Agronomic techniques of conservation are perhaps more in terms of preventing the start of the process of removal of soil particles, by the introduction of farming practices which will provide stable yields through time. Most conservation schemes combine both mechanical and agronomic techniques, and in many instances they are complementary. For example, gully erosion can be treated by mechnical means (e.g. a masonry drop structure with apron), cutting back the gully head to stable ground and planting it to grass and/or trees, and following a variety of agronomic measures above the gully to reverse practices which brought about sheet and then gully erosion in the first place.

Agronomic techniques were developed rather later than mechanical ones, following the identification of the importance of rainsplash as a major element in the erosion process. During the 1950s, and by the 1960s packages had been designed in the United States, which were then transferred to the very different physical and social conditions of the Third World. Since then international agencies such as ICRISAT and IITA have been undertaking further research on the more fundamental adaptations that now appear to be required (Hudson 1981). The techniques refer to new or modified agricultural practices to reduce soil degradation and erosion. Their recommendation to farmers is usually based upon an *ad hoc* or more systematic land capability assessment. This suggests a 'suitable' crop or land use and there may be very different weightings given by different land-users and government agents to what is regarded as suitable. New tillage practices are encouraged which seek to increase vegetative protection of the soil and reduce the effects of wind erosion and direct splashing of rain on the soil. Rainsplash washes elements downslope and reduces infiltration capacity by dispersing clay particles into the large interstices of the soil. New tillage practices also reduce loss of soil moisture to increase yields, build up humus formation, and encourage a net increase in the natural production of soil nutrients. To these ends, longer and new crop rotations can be introduced. These usually involve increasing plant diversity, contour ploughing, inter-cropping, minimum or no-tillage systems, and grass strips (with or without grass-covered storm channels or bunds, and with or without brushwood planting along the contours between strips). The development of forests, either by wholesale (re)planting of watersheds or in smaller areas such as gullys and riverbanks, in other small lots such as fuelwood plots, on terrace-backs and field boundaries for fuel and/or fodder, or in lines for the additional purpose of windbreaks, is also an important element in most conservation schemes. Pastures can be reseeded, and the quality of livestock increased by breeding programmes, increased veterinary services, improved supplementary feeds and fodder crops, all of which aim to increase yields and reduce the number of stock on degraded and eroded pastures. Many of these techniques of conservation have already been developed by cultivators and pastoralists themselves. There are many examples cited of local soil conservation: tie-ridging of the Kara and the Matengo pit system reported by Ranger (1971) for Zambia, contour ridging and inter-cropping in Tanzania (Stocking 1982), 'denshering' in western Cameroon (Kassapu 1979) and a variety of techniques discussed by Richards (1975). However, many of these locally developed techniques have been discontinued for reasons that are discussed later in Chapter 8.

There are also negative measures in the sense of restriction of land-uses, and exclusion of people and/or livestock from certain areas. Although 'conservation policies of land-use should be positive and encouraging, not restrictive' (Hudson 1971, 79), they frequently include these measures, which aim to stop certain agricultural or pastoral practices which are identified as harmful. These include shifting cultivation: for example, in Zambia where it is nationally outlawed, or in certain areas such as Sukumuland in Tanzania (Berry and Townshend 1973). Alternatively all cultivation can be banned from specific areas such as river banks (as in Kenya), watersheds or very steep slopes. Other restrictions apply to the felling or cropping of trees, even the collection of forest litter, either altogether (where forests are 'no-go' areas) or on a rotational basis (as suggested by Glover (1946) for hilly areas in the Punjab). Restriction of livestock with similar variations is also a frequent negative element in combinations of conservation techniques.

A similar classification of soil and pasture conservation is succinctly provided by FAO (1977b) and is reproduced in summary form on Figs. 3.1 and 3.2.

3. A review of conservation policies

Even a brief and superficial review of national soil and pasture conservation policies indicates that, with a few significant exceptions, their results fall far short of their intentions. While pilot projects and research station experimental plots often attract encouraging reviews and progress reports, programmes and policies almost universally have run into serious difficulties. The review here cannot claim to have gone into any great depth, but its purpose is correspondingly modest, and that is to provide some support to this generalisation. Maybe techniques of evaluation (important though they are) need not be sharp when faced with a strong weight of empirical and impressionist evidence of the failure of conservation efforts to halt the deteriorating environmental situation in substantial areas of the world. In a general sense the evidence examined in Chapter 2 (that in many parts of the world, soil erosion, floods, soil degradation and deforestation were probably getting worse) could be used in a crude sense to show that conservation policies have not succeeded. However, more evidence is needed about soil conservation programmes themselves. The sort of evidence for success that

Fig. 3.1   Techniques of soil conservation (from FAO 1977b)

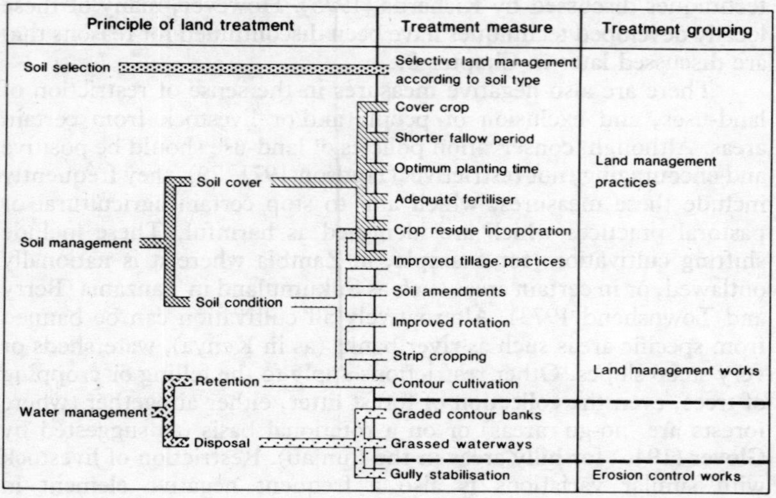

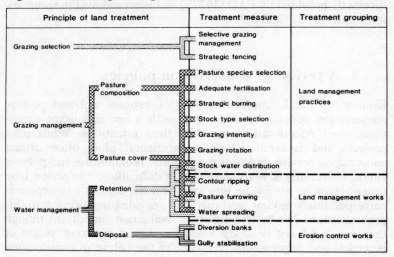

Fig. 3.2   Techniques of pasture conservation (from FAO 1977b)

one may look for is the fulfilment of the general objectives suggested by FAO (1976b) – that is, the increase of food production and standard of living of the people by means of soil and water conservation policies. Thus, a rapid review of conservation world-wide can be expected to provide evidence as to whether there exist programmes which *have* fulfilled these general objectives for substantial numbers of people (say, at least, tens of thousands), without insisting on precise standards of measurement. In this brief scanning of national efforts, the reasons for success or failure, the political-economic conditions under which they were made, and the technical details of conservation practices are only briefly mentioned. It is also assumed in this review that it is only officially sponsored programmes or policies that are being discussed. There are plenty of 'success stories' where governments have not intervened, or where it has not even been acknowledged that farmers and pastoralists have quietly got on with the business of conservation for themselves, and frequently provided sustainable surpluses for the market as well.

It is instructive to start with the USA, which in many senses is the country on which much of the methodological, legal, technical and institutional aspects of conservation policies of other countries have been modelled. A number of far-reaching evaluations have recently taken place, culminating in the Soil and Water Resources Conservation Act of 1977 (abbreviated to RCA). Rasmussen (1982:3) asks the question '... that might be raised today...is why erosion remains a severe problem after forty-five years of cooperative efforts by farmers and the federal government to solve it'. The title of Cater's article (1977), 'Soil erosion: the problem persists despite the billions spent on it', speaks for itself. A more equivocal comment that perhaps best sums up the evaluations that led to RCA has been made by Timmons (1980): 'While these measures [in soil conservation] have seldom attained goals, increases in soil erosion were minimised and occasionally soil losses reduced. Despite these efforts, erosion losses remain excessive and may have increased during the 1970s [and this trend] may continue indefinitely into the future.' However, the sophisticated methodology of land-use planning, as well as the policy instruments of the programme, have recently been shown to be very fragile and unable to prevent large areas of land previously left uncultivated (because of susceptibility to erosion), being once more put under the plough. Cook (1983) traces the impact of the decision to sell grain to the Soviet Union and the government's renewed intention to use food as a political weapon, upon the expansion of 27 million acres of arable land and a large increase in the annual rate of soil loss. In spite of a long-established programme, with the best technical expertise in the world, the programme over the years has achieved a very mixed success. Its importance in soil conservation practice is world-wide partly because much of the international literature on the subject in popular and scientific journals is written by Americans, and partly because of the American foreign aid programme in which conservation plays an important part. Large numbers of American aid personnel advise many of the governments of lesser developed countries. However the methodologies in assessing soil erosion hazard, as well as policy instruments, may well not be applicable outside the United States.

Moving now to lesser developed countries, a general review of conservation programmes of Algeria, Tunisia and Morocco (Bensalem in FAO 1977a: 159) concludes that 'soil conservation practices in North Africa have been focussed mainly on corrective mechanical measures, which in most cases have been insufficient or not suitably adapted to counter erosion problems' He goes on, like so many other commentators, about the lack of applied knowledge of both the appropriate technical options (e.g. the successional capacity of the native plant cover which can be used to rehabilitate degraded land), but also, very significantly, 'there are soil conservation and dune afforestation schemes dating back thirty years, some succeeding and others failing, but without explanation of why' (p.159). Another older FAO report (1966) suggests that a total lack of local participation in conservation measures was the major reason for widespread failure. Heusch (1981: 423), commenting on conservation in the Rif mountains, Morocco, comes to the conclusion that:

> soil erosion in the Rif is the by-product of the submission of highlanders under more dynamic groups of high living standards. This leads to a neglect of conservation management ... the soil conservation expert is thus requested to propose technical and educational solutions to socio-political problems, whilst the sociologist is given the task of obtaining the approval of the population for measures taken without its advice and real consent.

This quotation has many interesting strands, some of which are explored later on in this book. Nigeria's soil erosion problem has been recognised as extremely serious but conservation has only become a national concern during the Third National Plan (1975-80). Olayide and Falusi (1977: 115) comment that 'past government efforts in soil conservation have been very meagre and have had little impact in solving the erosion problems'.

Moving into Central and Southern Africa, Zambia's postcolonial conservation programme is seen by Robinson (1978b) to have been adversely affected by the preceding oppressive colonial efforts at conservation. Politicians are reluctant to popularise or enforce measures against which many campaigned prior to Independence and 'the majority of farmers remain unaware of the causes of erosion and unconvinced of the need for or value of conservation measures' (ibid.: 28).

India's long-established conservation and forestry programmes are well documented in terms of plans and normative statements about the future, but lack evaluations. Agarawal (1980: 109) points to failures of social forestry programmes also documented by Alvares (1982). Agarawal also points out other failures of forestry and fuelwood projects in Java (p.105), and Upper Volta (p.166), mainly blaming a lack of a community of local interests. Bali (1974) talks of the need for reorientation of the whole of conservation policy in India, while Ranganathan (1978) attributes failures of past efforts to inter-departmental rivalry and lack of extension. Also Patnaik (1975), Parikh (1977), Gribbin (1982), and Kulkarni (1983) leave little doubt of the glaring failures of Indian conservation. Perhaps the two most comprehensive critical reviews of the Indian soil conservation are the Centre for Science and Environment's *The State of the Indian Environment* (1982) and Cahill's report for the FAO (1981). The former concentrates on what is actually happening, comprehensively outlining the growing problems of deforestation, desertification, soil erosion, flooding, siltation, as well as a decline in standards in health, housing and industrial pollution control. From both this report and that of Cahill (ibid.) it is quite clear that the problem is severe and getting worse; and that the national soil conservation programme is not dealing with it:

> ...looking into the future, what is important is not so much the *extent* but the *rate* of progress with soil conservation. At a completion rate of one percent per annum the total task would take 100 years. The actual rate of completion per annum over the last 27 years has been less than one half of one percent...thus even at several times the projected rate of progress, the great bulk of the conservation task will be incomplete at a time when the most vigorous additional demand is placed on land from massive upsurge of population. (Cahill 1981: 2)

Thailand's efforts to stem serious erosion problems resulting from the extension of cultivated land as a result of population growth and an expansion of a market for commercial crops, and large-scale destruction of forests by logging companies and (to a lesser extent) local cultivators, are reported to be largely ineffective (Kilakuldilok 1981), El-Swaify et al. (1982). Kilakuldilok attributes this to a familiar list of reasons—institutional weaknesss in all aspects of conservation, lack of political will, rapid population growth, and lack of a coherent long-term policy. There have been a small number of fairly successful conservation projects, notably Chang Mei, and at Ma Sae (UNEP 1979: 83), but they do not materially detract from the view that Thai conservation policy is not succeeding in reversing national trends of rapid forest clearance and increasing erosion hazards. Malaysia's efforts are also judged to be inadequate (MacAndrews and Sien, eds. 1979: 30; and by Hamzah bin Abdul Majid, also in MacAndrews and Sien, eds.: 54).

South Korea's reafforestation and erosion control programme is one of the few that has enjoyed considerable success. Although there had been a long established tradition of community and private forestry schemes, it was not until the launching of the *Saemol Undong* (or New Community Movement) in 1970 that it was possible to conceive of a nation-wide forestry policy and programme. A ten year Forestry Development Plan was initiated in 1973, and it was organised largely within the institutions and objectives of *Saemol Undong*, which emphasised effectively a broad-based approach to achieving rural welfare, an incremental step-by-step planning by villagers for their own village with an emphasis on short-term gains in income, and adequate government back-up services. Physical targets were mostly achieved four or five years ahead of schedule (FAO 1982a: 67), although there were shifts in the emphasis between sub-targets (e.g. from fruit and nut trees to long rotation species). The whole programme consisted of reforestation, nursery operations, tending operations, and erosion control (hillside treatment and sand-dune fixation). The political will came at a critical time when it was very necessary to keep the food costs component of the total wage bill (including the foreign exchange component) as low as possible to allow rapid and profitable industrialisation. Since most of the surplus food was from the lowlands with irrigation water originating in the highlands, widespread deforestation in the latter areas had had very serious effects upon the ability of South Korea to feed itself. In addition large rural-urban income differentials and massive urban migration contributed to a very unstable situation from about 1962-70. Thus, in South Korea there were exceptional circumstances in the political economy of the country that led to a highly vigorous programme of rural uplift, of which conservation was only one part. For a more general discussion of South Korea's development, see Brandt (1978) and ESCAP (1979).

China's conservation efforts have received a good deal of attention and it has been difficult to sort out myth from reality. This was partly due to the West's indifference to information gathering until recently, the lack of scientific and cultural exchanges until China became a member of the UN in 1971, and extreme xenophobia on the part of the Chinese government engendered by the Cultural Revolution. It is undoubtedly one of the major conservation programmes of the world. Very favourable accounts by Sandbach (1980) and Qu Geping (1980) have recently been qualified by Howard (1981) for the highly vulnerable loess soils; some reservations and the need for further mobilisation of the masses as expressed by Dequi *et al.* for the Winding River Valley (1982); and a number of more damning commentaries on China's forest policies (Eckholm 1976, Smil 1979, Delfs 1982). Undoubtedly there have been numerous policy turnabouts, most noticeably at the time of the Cultural Revolution when forestry resources were misused for unsuitable land reclamation (Gustafsson, undated), which in certain particularly vulnerable areas may have exacerbated the destructive floods of autumn 1981. But the combination of the daunting task of national reconstruction after the Communist victory in 1949, the neglect of conservation for centuries, and a variety of purely physical factors making for great erosion hazards, must also be taken into account. Thus, as in the case of the USA, the size and complexity of the programmes, mixed successes, and the possibility that programmes have significantly reduced soil losses, or even the rate of decline rather than improved yields, point to the need for a longer and more sophisticated assessment than is given here.

In Latin America, the state of soil erosion programmes is even worse (Preston 1980). Freeman (1980) estimates that 80 per cent of Bolivia's farmlands and grazing areas are eroded in varying degrees, and rangelands are degraded down to 10 per cent of their original productivity. He estimates that between 35–41 per cent of all useful land is affected by erosion. In response to this situation 'the national government has made no response to the problem of erosion, and none is planned'! Paraguay does not even have a department or other state institution which is supposed to initiate soil conservation and range management. In the case of Brazil, evidence (presented elsewhere in this book) points to widespread erosion, particularly in the *Selva* and in the *Serrao* of north-central and north-eastern Brazil. In one consultant's report it is stated 'Brazil has no institutions that are capable of undertaking the task of achieving soil conservation and preventing land degradation.' At the Federal level there is an Office of the Co-ordination of Soil and Water Conservation, but it has a tiny staff and its position within the Ministry of Agriculture does not enable it to have much influence on land-use decisions. At State level, there is simply no administration at all to carry out conservation, although at present State Soil Conservation Commissions are in the process of being formed. It comes as no surprise that Brazil's rural credit policy actually encourages farmers to follow bad farming practices and there is no countervailing influence to change it. The activities of both large transnational and national companies as well as small-farmer settlers on the northern frontiers of settlement have often been disastrous to the long-term productivity of the land, but have gone on unchecked (Barbira-Scazzocchio 1980).

This very brief review of about ten conservation programmes cannot make the case that all national policies fail. The authors and quotations mentioned, however, are representative of a wider literature of evaluation, but it is the absence of favourable comments which supports the view that this small sample is representative of most other lesser developed countries.

Chapter 4

Why do policies usually fail?

1. Introduction

In this chapter we will examine various explanations of why most soil conservation policies do not work. An analysis of the mode of explanation itself is helpful since it makes explicit many unstated assumptions on which these explanations rest. The conclusion will be that these explanations have recently developed in promising directions. However, most still fail to state explicitly their political value judgements and are unaware of their ideology.

One of these promising directions has been the realisation that conservation is as much about social processes as physical ones, and that the major constraints are not technical (in the agricultural engineering sense), but social. Here are four illustrative quotations from the recent literature:

> Land-use patterns are an expression of deep political, economic and cultural structures; they do not change overnight when an ecologist or forester sounds the alarm that a country is losing its resource base. (Eckholm 1978: 167)

> The theme of this paper is that knowledge of soil conservation techniques should be improved, but the most pressing need is to put into practice what is already known. Past experience of implementing soil conservation programmes has been disappointing and the paper tries to identify the more important reasons. (Hudson 1981: 15)

> While socio-economic criteria may often exist in the minds of the planners, there is an alarming failure to relate social and economic needs to acceptable forms of conservation. In the final analysis, if the small farmer and peasant of the Third World cannot or will not take conservation to heart, no amount of model-building building, empirical plot studies, erosion risk assessment or legislation will result in the preservation of the soil's resources. (Stocking 1981: 377)

> What should be said is that there is still a most definite need for research in some neglected, technical areas, but more particularly in integrating the social and economic factors as they occur in a local situation. Without this integration the results of existing technical research cannot be applied in less than the present high risk of failure. (MAB 1979: 9)

However, these 'social factors' are not subjected to the same systematic analysis as the technical ones, and they remain unconnected, irksome and somehow immovable under prevailing analytical frameworks. The analysis which follows in the succeeding five chapters attempts to show that they *are* irksome; and indeed many are immovable without other deep-seated changes in society; and that these cannot be advocated or struggled for on the grounds of soil conservation alone.

Our starting point lies in the identification of the social factors which help to shape the *perception* of the problem of soil erosion by those who attempt to intervene in land-use decisions through government action. Most of the attempts to include social factors in the study of erosion and conservation, as the four illustrative quotations show, imply that the social problems start with 'them', the land-users themselves. Here we argue that a comprehensive analysis should broaden the scope of the analysis to include conservationists and governments. Many readers (and this author too) might therefore find themselves as objects of analysis. The analysis of soil erosion includes those with political power and asks embarrassing questions about their own view of the problem, and from where it derived. It points out that peasants and pastoralists may seem tiresome and contrary, but this is an interpretation of political-economic relations which envelops the perceiver too. When official conservation often does not work, part of the problem may lie in the assumptions which the conservationist, government servant or politician may make about the cause of failure. These causes range from backwardness of peasants, lawlessness and lack of discipline by land-users, and inefficient implementation procedures, to lack of 'political will'. Behind each cause lie assumptions about the way in which society does and should operate.

Appealing to rather different sentiments, there are strong grounds for arguing that scientists should take social responsibility for the technology they develop. Some scientists would still maintain that science is neutral, and it is the politicians who select and implement their inventions in particular ways and who should therefore shoulder the blame for undesirable consequences. In a less dramatic and analogous fashion, soil conservationists of all kinds should be able to take a social view of conservation techniques. Bunting (1978) in more general terms disclaims this responsibility and states that with the introduction of high-yielding paddy it was to be expected there were social problems but it was up to the politicians to 'do something about it'. The view that, somehow, science (here, agricultural technology) is neutral and it is the fault of 'irrational' politicians or pressure groups is a widely-held one – even if a little sheepishly.

Oasa and Jennings (1982) go further and show the disdain and hostility that the International Center for the Improvement of Maize and Wheat (CIMMYT), and the International Rice Reserarch Institute (IRRI) showed for the comments and criticisms of social scientists about the social implications of the technology that these institutions were developing. These criticisms were ignored (although delivered by persons of world-wide reputation such as Carl Sauer), received a hostile and defensive reaction, or were absorbed by transforming them into a technical issue – rather than facing them as a social and political one. This last point is one which is developed here in relation to soil conservation. The charge of political bias laid at the door of critics, and the claim that their own scientific activities were neutral, leaves only one way out – to make out that the problem lies in technical factors of production that constrain maximum yield performance in maize, and *not* in the interrelatioinship between social conditions of inequality and the new technologies which CIMMYT and IRRI were developing.

The next section discusses a model of soil conservation which derived from the perceptions and objective political economic conditions of the 'colonial period', stretching from about 1880–1960. Since that period has passed, and we can benefit from hindsight and a somewhat altered world view, the social factors which moulded this model can be more readily recognised. During the past twenty years or so, many soil conservation programmes in lesser developed countries have been initiated, financed and often partly staffed by foreign aid institutions, particularly multilateral agencies such as the Food and Agricultural Organisation, and the World Bank, and also the biggest bilateral agency, USAID. These institutions are very different from the colonial ones of the past, but none the less many of the assumptions that lie behind today's conservation policies and projects remain intact. Furthermore, much of the writing and even foreign staff in conservation institutions are derived directly from colonial administrations. Foreign aid is so important in conservation policy-making in lesser developed countries because, with a few exceptions, most newly independent lesser developed countries had either no official conservation organisation whatsoever (as in Latin America with the exception of a couple of Southern Brazilian states) or had experienced the colonial model mentioned above, principally in Africa, South and South-east Asia. For both reasons, there was either virtually no official concern about soil erosion or a reaction against enforced policies which were firmly linked to the colonial regime. Under those circumstances, foreign aid agencies saw a vacuum which they felt themselves able to fill. Because of the deeply embedded social issues in soil erosion and conservation and the particular political-economic relations between donor agents and lesser developed countries, new problems tended to arise, and these are discussed in section 3.

2. The classic or colonial approach to erosion and conservation

It would be an unjust and simplistic generalisation to claim that most conservation policies and official reactions to soil erosion followed a single set of assumptions. However, it is possible to identify a syndrome of implicit assumptions in many lesser developed countries' conservation policies and in more academic and general commentaries on the subject, and to follow them through to particular elements in explicit policy. Few actual conservation policies contain all these elements and it is difficult to attribute them in every case to implicit value judgements. A 'classic approach' implies a static, long-established, well-recognised and highly valued approach on which particular policies in the future may be based. Both the theory and practice of soil conservation are in a process of change, albeit rather slowly, and in the opinion of the author often in directions that are not very promising. In spite of this caveat, the implicit assumptions on which policies are based have only slowly evolved from a colonial, Euro-centric and messianic intellectual frame of reference which has endured the waning of empire and the regaining of political independence of most former colonies. This is described below. The classic approach identifies four major problems in rectifying soil erosion.

social factors acknowledged & barred?

(i) Environmental problem/solutions

The problem is identified as primarily an environmental problem with environmental solutions. The opening chapter of this book has put an alternative view, and in spite of an awareness of the importance of 'social factors' in environmental degradation, there is very little evidence of effectively implemented policies. If soil erosion is principally identified as an *environmental* problem, then the 'degrees of freedom', and policy options which derive from concomitant changes in the *social* sphere are thereby closed. Therefore there is a logical tendency to preclude from view the *social* reasons why people are using the land in such a way as to cause excessive soil erosion, and so most conservation policies do not address them at all. Problems of a lack of alternative sources of fuelwood for poor people, unequal landholdings, political constraints placed upon the pastures of nomads, poor prices imposed by the state for the produce of rural areas and so forth, tend to lie outside the terms of reference of most conservation policies. When purely environmentally-dictated policies (such as forest closure or forcible destocking of pastures) do not work because the social context has been analysed insufficiently, force is often contemplated (or, in administrative language, better and tougher implementation and environmental protection). This frequently ends up with the state attempting to protect the environment from the majority of the people who use it. For example, the Agricultural Department of what was then British Kenya, when faced with the Wakamba who had marched on Government House as a protest against compulsory destocking in 1938, said:

> ...unless some pressure is applied to urge improved methods and practices, and unless such pressure is continuously applied...it will not be possible to save the fertile areas of Kenya from deterioration...without the application of compulsion under legislation to enforce improved agricultural practices. (Clayton E. (1964) citing the Agricultural Department of Kenya)

A more up-to-date but similar sentiment advocating a tough approach:

> Admonition alone does not appear to be enough. A strict legal regime of soil conservation must be established. (MacAndrews and Chia Lin Sien 1979:30)

Policy 'solutions' of this kind tended to favour erosion works (often paid for by labour inputs on the part of farmers themselves), terracing, reafforestation and pasture closure, implemented by compulsory destocking or exclusion from certain vulnerable areas. They were frequently imposed by colonial regimes using force where necessary. In colonial Ruanda-Burundi, several weeks of free labour had to be given each year by the *commune* to build up terraces, bunds and other erosion works. The *paysannat* system in the (then) Belgian Congo had similar compulsive elements (Dumont 1970: 36).

Mismanagement of the environment

The problem is blamed on the land users themselves, who are seen to have mismanaged the environment because of lack of environmental awareness, ignorance, apathy and just plain laziness. The more pejorative epithets have disappeared in the post-colonial literature, except for one or two extreme examples like Kon Muang Nan (1978) who holds the view that the shifting cultivation practised by tribal people in northern Thailand is the most dangerous national problem and one which undermines national security. The practice should be eliminated or the people expelled from Thailand altogether, he claims. Here are two more highly pejorative explanations of soil erosion:

> The natives are simply running wild and ploughing up large areas of land every second year. Not only are they ruining good soil through bad farming, but destroying all the valuable timber. This is a country that can support a lot more natives if properly settled. The main offenders here are non-indigenous natives who have cleared areas and acres of the best soil, farming it very poorly and

working it as though it were a private farm...there is no rotation of crops and no manure applied which means he will soon be looking for another 100 acres of good soil to ruin. (Inspection Report to Chief Native Commissioner, Southern Rhodesia, 1938, in Stocking 1978b)

One might ask why the non-indigenous whites who ran private farms (and thereby improperly 'unsettled' other natives) are assumed to 'use' rather than 'destroy' all the valuable timber.

Farmers are well known for their conservatism. The African agriculturalist is no exception and is very tenacious of the customs and methods practised by his forefathers...the poor farming methods and soil depleting practices prevalent among peasant cultivators stem from ignorance, custom and lethargy... the main obstacle to overcome is the native's lack of understanding of the need for the prevention of soil erosion. (Clayton 1964)

Here are three more restrained, post-colonial examples, but none the less with the same implication:

Inadequate knowledge of basic factors resulting in excessive or untimely tillage, improper implements, poor use of the right equipment, burning of crop residues, and excessive livestock grazing have all contributed to erosion on cultivated land...(FAO 1960: 1), leading to such a condition which should not happen if all people who use and work the land could have the advantage of scientific knowledge relating to soil and water conservation. As quickly as possible all countries should adopt an enlightened land-use policy and provide means of carrying it out. (FAO 1965: 13)

It is basically a problem of the misuse of land...particularly in pastoral areas, much of the problem results from the customs, value systems and attitudes of the people concerning grazing lands and livestock, together with the lack of government mechanisms for effective control. (FAO 1980: 56)

The most recent and important example comes from the proposed Forest Bill of India which is at the time of writing about to come before Parliament. It illustrates many points in this chapter, but above all the point about blaming forest users. The bill identifies much of the blame for the tremendously high rate of deforestation in India as belonging to the *adivasis* or tribal peoples, who have been systematically marginalised and impoverished for over a thousand years, and who have now retreated into the mountains and remaining forests of central India (a close-up case study is provided of a tribal group, the Sora, in sect. 7.4). The Bill suggests prevention of the *adivasis* using the forest, very heavy fines (double if infringements are perpetrated at night) and powers of arrest without warrant by forest officers. Ample evidence exists of the real destroyers of the forests being private contractors who supply the timber demands of industry (Centre for Science and Environment 1982; Gribbin, 1982).

The policy implications of this view of the problem are spread across a continuum from agricultural extension to educate farmers into new and more modern farming practices at the one end, to compulsory erosion works and outlawing practices at the other. Again the root of the problem was perceived to be the cultivators or pastoralists themselves, and that blame originated from land use by them under their own free will. If 'they' can be made to change by education or compulsion, then the problem of soil erosion and pasture degradation would be solved.

(iii) Overpopulation

Another problem identified in the classic approach is that of overpopulation.

> Any solution to the problem of (environmental) deterioration must first cope with the basic cause; overpopulation. (FAO 1980: 57)

The whole issue of the politics of population growth rates in relation to natural resources and production technology is a highly complex one, and not discussed in any great detail here. However, it is enough to point out that there are similarities between the 'overpopulation' view and the view that farmers and pastoralists should be educated out of their ignorant, lethargic and traditional ways. Both identify the cause of the problem as beginning and ending with the land-users themselves ... they should change their habits of production and reproduction, and the problem of soil erosion would be largely solved. In some policies the two problems are seen to be combined where the systems of farming collectively called shifting cultivation have reached a point where their carrying capacity has been exceeded, and fallow periods have become so short that degradation and erosion have set in (Allan 1967). The policy implication here is that family planning programmes are linked strongly with conservation programmes (as in Kenya or in India in Sanjay Gandhi's Four (and later Five) Point Programme which included both family planning and reafforestation).

(iv) Involvement in the market economy.

The last problem identified in the classic or colonial model is that cultivators and pastoralists who cause soil erosion are insufficiently involved in the market economy. Involvement in the production of surpluses for sale in the market implies modern methods of cultivation and improved productivity, so alleviating the 'population problem', and the awareness of financial inducements, and incentive to undertake soil conserving agronomic practices and/or erosion works.

The policy implication of this view is a programme to help those farmers who can help themselves to grow cash crops.

A quotation (provided by Randall Baker, private communication) from a recent Australian funded cattle ranching scheme in Fiji illustrates both the unquestioned assumption that development must imply modern commercial development and the disparaging attitude towards existing social and economic organisation:

> The Fijian traditional communal system of livelihood has a tendency to restrict initiative for commercial expansion and development so that there is a need to modify commercial values to meet with the demands of modern commercialism. This, in a nutshell, is what the Yalavou project sets out to do. (Fiji Government 1982: 7)

A case study of Kenya (Baker 1981) admirably illustrates many of these points, and Fig. 4.1 from Baker is reproduced summarising many of the points made above.

The political and economic origins of this approach to soil conservation go a long way to explaining it. It was an approach not of the land-users themselves but of their rulers, and therefore it is necessary to discuss briefly what the colonial rulers' political and economic interests were, and in what way they related to the people of colonised areas and to the natural resources they found there.

Fig. 4.1 The technocratic perception: environmental protection (from Baker 1981)

PROBLEM	SYMPTOMS	CAUSES	SOLUTIONS	CONSEQUENCES
Kenya has an → Environmental Crisis	Desertification → Deforestation	Overpopulation → {overgrazing } {overcultivation}	Family planning →	Lack of response
	Soil erosion Catchment loss Silting	Ignorance → {tradition {culture {inapp. practices	Education → Change attitudes Demonstrate → new ideas	Inappropriate knowledge = frustration Short-term palliatives
	Decline of rivers Decline of food production	Lack of → environmental awareness Inadequate → legislation	Environmental → education and EIA Tougher legislation →	Rationalising oppression Oppression and polarisation Protect environment against people
		Institutional → weaknesses	Integration Min. of the → Environment	New and more efficient ways of avoiding problem

Considerably more detail is given in Chapter 7, and only an outline of the major points as they contribute to understanding of this approach to soil erosion appears here. It is interesting to note that soil conservation as a national policy, supported by the various apparatus of state (agricultural extension, the law, and law enforcement agencies) was largely an African colonial phenomenon (French, Belgian and British), with a smaller constituent from colonial India and the Dutch East Indies. Therefore the fact that most of the illustrative material comes from Africa, and not from Latin America, is not by chance.

First of all, although colonial policies varied considerably through space and time, one of the chief resources for colonial regimes was land. Either local cultivators and pastoralists were cleared from the land completely and it was reserved for European immigrants, or in some cases (like Uganda and larger parts of Tanganyika) local cultivators were encouraged, or forced, to grow crops for sale either as export crops or as food crops to feed the mining workforces in southern Africa (for longer accounts of the processes of settler immigration and colonial policies, see Rodney 1972; Palmer & Parsons 1977). One of the major implications was extreme overcrowding of the local population in terms of their usually extensive farming systems. Usually too, the European took the best land as in Algeria, for example, where 20,000 European farms were established on a third of the cultivable land (about 2.5m. ha.) while some 630,000 Algerian peasants were confined to 5m. ha. of dry, hilly and thinly soiled areas (Stewart 1975). Similar displacements occurred and Africans were confined to reserves in Kenya, or to tribal trust lands in Zimbabwe (Southern Rhodesia) and at the present time in the Bantustans by the Republic of South Africa. A similar displacement occurred much earlier in the New World where the *enconomienda* system was established in Latin America and a plantation economy in the Caribbean islands, both of which suddenly displaced local cultivators. The existing cultivation systems often caused serious erosion as a result of the 'instant overpopulation' which ensued. Similar fates befell pastoralists where large areas of seasonal pasture were denied them as in Kenya in both the Masai territory and Karamajong. By assuming the political economic circumstances of this displacement as given, of course it seemed natural to the colonial administration that the condition of environmental deterioration was the fault of the cultivators themselves.

Second, it was in the economic interests of the colonial (and home) administrations to persuade or force the cultivator to grow produce for the market either because it increased the cultivator's taxability and therefore he/she could contribute to the infrastructural administrative and policing costs of the colony; or, where settlers would or could not do the job, it would provide a source of cheap food for the mining workers, particularly in southern Africa. Hence the motive for commercialisation of the peasant created a convergence between a philanthropic desire to improve the lot of the peasant and to serve the economic interests of mining companies and colonial administrators. Many other contemporary examples of both national as well as international institutional policy in which the market is seen as the only means to improve peasants' and pastoralists' incomes and to increase their contribution to the national exchequer can be found in Heyer, Roberts and Williams (1981).

The introduction of new cash crops such as groundnuts (particularly in the Savannah areas of Africa south of the Sahara), coffee, cotton and maize both by European settlers and to local cultivators throughout eastern and southern Africa, was strongly supported by all colonial administrations. Research stations undertook research for the larger (white) commercial farmers and most African cultivators were actively discouraged from producing in competition with settler farmers (Brett 1973). However, the introduction of these crops often had serious consequences for soil conservation such as the cultivation of groundnuts in Mali and Niger (Franke and Chasin 1980), and pure stand maize cultivation and cotton. The repercussions of the introduction of cash crops have been analysed in detail in postcolonial times (Dinham & Hines 1983), particularly upon a reduced capacity of households to grow enough food for themselves, upon nutrition, and upon their ability to withstand drought and other, socially-induced, disasters (Copans 1975; O'Keefe et al. 1977).

Of course, a set of policies on soil conservation was not simply determined by a few major economic and political interests of colonial powers alone. For one thing, they varied enormously from place to place (e.g. the distinctions between the policies of the colonial administrations of Kenya and neighbouring Uganda, Brett 1973; Leys 1975), and between Anglophone and Francophone Africa. They varied through time, and there were also important contradictions between local governors who understood well these deleterious effects upon cultivators and pastoralists, and their paymasters both in the colony and metropole (Brett, ibid. Chapters 6 and 7). Last, but not least, there was an attitude of mind of the rulers towards the ruled, varying between paternalism through uncomprehending distaste to outright racism. Africans received laws, fines and even imprisonment for certain agricultural and pastoral practices (and the other restrictions which enabled Europeans to realise their economic interests of the day), while European settlers received physical infrastructure, loans, and an agricultural extension service backed by research stations which researched commercial crops for commercial farmers.

What evidence is there that these overt policies and covert assumptions have outlived the political and economic structures which brought them into being? After all, most countries which were colonies in the past, have regained their independence. There are two lines of argument here. The first is that the conservation policies themselves still have many of the assumptions of the colonial model. Some post-colonial examples have already been given, and more appear in Chapter 7. The second argument is that many structures of appropriation of surpluses from peasantries and pastoralists which were established during the late colonial period still exist. While political independence is a contemporary and important fact of life in all but a handful of countries, the economic incorporation of their economies into the world economy continues and deepens. The aspects of this process which affect land-users are discussed in detail in Chapter 7. Briefly, the incorporation of an economy into the world economic system involves an extension of the market mechanism between producers and consumers in the lesser developed country with other countries or transnational companies, and thereby a closer political relationship between other trading partners and national or international financial institutions. Incorporation is both affected by the effects of the internal class structure of the lesser developed country and, as it deepens, becomes a determining process of social development or retrogression. Although the approach of this book makes its political stance clear, it is not its intention to undertake a wide-ranging critique of various models of social change. It is not necessary for readers to agree completely with the ideology of the approach to arrive at some of the same conclusions about soil erosion as found here. For example, if readers agree that incorporation is the only feasible and desirable route to social development (e.g. Bauer 1976: 66f.), then soil erosion has to be counted as an inevitable cost. However, the reader could still maintain an internally consistent position by advocating 'incorporation' as the only feasible route to modernisation and development, at the same time as acknowledging the linkages between these processes and soil erosion. To expose the contradictions of the world economic system by identifying soil erosion in lesser developed countries is like expecting the tail to wag the dog. Soil erosion is one of its many contradictions, and for reasons amply discussed in this book, it is one of the most diffuse and complex areas of analysis. It is not one to rally the world's peasantries, pastoralists and other land-users to change the social conditions which bring about soil erosion in the first place. We return to this point in Chapter 9.2.

3. Foreign aid and conservation programmes

Many conservation programmes in lesser developed countries have been initiated, financed and partly staffed by foreign aid donors, both multilateral and bilateral (of which by far the most important financially and intellectually has been USAID). Given the fact that soil conservation as a government policy was a colonial phenomenon, it is perhaps not surprising that in post-colonial Africa at least, foreign aid tended to move into the vacuum left by the colonial administration. Also, the USA exported its long experience with trying to deal with its own soil erosion problems, as a part of foreign policy to its sphere of influence in Latin America in the 1950s, where there had been an almost complete lack of government concern over soil erosion. Foreign aid has received a number of critical analyses, the best known of which is by Lappé et al. (1980) and, in relation to the activities of the World Bank, Payer (1982), and these will not be reviewed here. In a sentence the kernel of the criticism is that foreign aid tends to serve narrow donor interests and those of the classes in recipient countries that will most benefit; that it incorporates the lesser developed country into dependent relations with industrialised nations; and that through a series of bureaucratic filters, special interests and general bungling, frequently does harm not good. Soil conservation programmes, by the nature of their objectives and the distribution of personal rewards (which are usually meagre since there is little money to be made out of them) are immune from many of the more trenchant criticisms of foreign aid policies and programmes. However, the fact that soil conservation and other related programmes are delivered through foreign aid institutions poses additional problems for their success. The elements which are essential in most conservation policies are often contradicted by the limitations and objectives of foreign aid. Table 4.1 below summarises some of these contradictions.

The results of these contradictions tend to follow a pattern. First of all there is the project document of which the format and content are so often quite unsuited to tackling the formidable tasks required for a successful conservation programme. Typically there is a short anodyne introduction to the economy of the recipient nation, emphasising the need for conservation and the nation's laudable attempts to meet it. There is seldom any serious political analysis either of the ruling class(es) and how they express their interests through the state and its institutions or of 'civil society' where the conservation is supposed to happen. There are obvious reasons why not, but its absence contributes to a continuation of failed projects because authors make heroic assumptions and quantify them to produce fanciful targets. Political-economic analysis could help professional economists both by providing a critique of their

Table 4.1   Contradictions between foregin aid and soil conservation programmes

Essential elements in conservation programmes	Objectives and limitations of foreign aid projects programmes	Result
Long maturing	Measurable benefits within three to five years	Emphasis on short-term and often peripheral objectives to soil conservation
Diverse, timely and highly coordinated inputs	Inability to deal with a large number of line ministries	Either disorganisation, or attempts to set up independent foreign-staffed implementation agencies
Outputs diffuse, diverse and difficult to quantify	Quantifiable benefits need to be predicted at proposal stage for purposes of justification	Concentration on physical and often less important objectives of conservation
Implementation deeply involved in sensitive political issues	No overt interference in internal political affairs of recipient country	Acute problems of implementation
In-depth analysis of political economic circumstances	Short-term consultants with the necessity to be tactful on political feasibility	Project documents full of rhetoric and technical details
Sustained political will at central government level	Short-term consultancies (1-3 years usual)	Uneven and uncertain back-up of project/programme after aid finishes

ideology, and also in assessing what is feasible, who will benefit and lose from their policies, what will tend to be done or left undone in the project and so on. The project document continues with a good deal of detail about the erosion works, areas to be covered by different treatments, and the expected benefits. An economic appraisal is provided usually on a before-and-after, or with-and-without the project basis. Here, most of all in the project appraisal, does quantification mystify and legitimate in the name of objective economic analysis. Busy bureaucrats, accountants, division chiefs and so on can always check the budget to the last dollar, suggest improvements to the 'organigram' of the project, add a foreign-training component and so on. On the other hand it is a lengthy task, based on shrewd local knowledge and acute economic and political-economic judgement, to appraise the appraisal. So the organisation of personnel, budget, physical specifications of build-ings, office equipment, etc. take up the rest of the project document and its appendices. There is little time, and little scope for an in-depth analysis of the political-economic context of the conserva-tion programme. Lest this account seem unduly cynical, it must be added that the individuals involved are subordinate to the functional constraints of their job. They are not paid to raise unnecessary dust in the formulation of a project; they are paid to design a project to spend earmarked funds in a way that they see as 'best'. Even those who are acutely aware of these constraints have to be content with doing their job to the best of their ability, although they may be aware of the probable shortfall in impact of the whole project.

The project is then signed by representatives of the donor and recipient country and implementation starts. A search through a considerable number of mid-term and final reviews and evaluations of conservation projects by the author produced an interesting pattern. However it is a very difficult and time-consuming operation to read through the evaluations and to compare them to the original project document. Here are some of the findings from a search through project documents and reviews:

(a) Physical achievements with an exact specific output are usually reached, such as buildings, most erosion control works, training programmes, aerial photography, mapping sub-projects and so on.

(b) The build-up of trained personnel and institutional development also usually goes according to schedule.

(c) The development of agronomic research into soil conserving practices also (in four cases where the search discovered projects with this component) seems to have occurred at the research station, and on the land of a handful of farmers. Outreach to land-users in the project area was in all cases 'slower than expected', or seemed to have been forgotten altogether in the evaluation.

(d) Where conservation was part of a wider package of rural development programmes (e.g. in integrated rural development projects), the physical outputs were achieved on target such as construction of storage facilities, target areas planted under some improved varieties of crops, the planting of trees in big plantations, the establishment of sawmills, agro-processing plants, mills and packing stations. However, other more ambitious targets such as reducing the rate of soil erosion, improving the offtake of pastures and the diffusion of soil conserving land management practices almost always were not achieved. Benefits were distributed socially and spatially less widely than anticipated.

This summary of findings does not refer to all foreign aid projects. It is the combination of the particular elements required by successful conservation projects with the particular political, economic and logistic constraints under which foreign aid has to operate, that limit the fulfilment of objectives. The problems discussed in this section are not necessarily separate from those outlined as deriving from the colonial model, indeed they frequently overlap. When they do, the results are all the more dismal.

4. Some explanations for failure, and some reassessments

Chapter 3 has suggested that many soil conservation policies fail, and section 2 of this chapter has identified an historical model of the conceptualisation of the soil erosion problem and its policy solutions. Although many elements of that model still exist today, there is also quite widespread dissatisfaction with it, and there has been some attempt to find out why, and to modify these elements. Although this effort has been piecemeal and has occurred in widely differing contexts, the direction in which it is moving can be identified. Still, many of the most fundamental 'social factors' behind the conceptualisation of the causes of soil erosion and the role of government have still not been analysed coherently, and remain at best labelled and hidden in a number of black boxes. Here are some of the reasons that are commonly found in contemporary documentation for failures in conservation programmes.

(a) Conservation techniques do not conserve soil in practice because of technical failures through inadequate or misapplied research. These range from the collapse of concrete-walled and widened terraces constructed by the Israelis in occupied territories (Bunyard 1980) to repeated failure of inflexible concrete check dams (nowadays frequently replaced by gabion), breached bunds, inadequate storm channels and inadequate knowledge of the requirements and performance of grasses and trees. Other examples include Nobe and Seckler for Lesotho (1979); Robinson for Zambia

(1978a 1978b); Coulson (1981); Rapp (1975) and Rapp et al. (1973).

(b) Conservation techniques do not fit into agricultural and pastoral practices and therefore are not applied by farmers or pastoralists. An extended example of local conservation techniques being neglected or even discouraged and government initiated ones being encouraged (and later on forced upon the local population) is given by Berry and Townshend (1973). Table 4.2, adapted and added to the original (ibid: 246) indicates the techniques, their origin and their acceptance. One or two extra and up-dated pieces of information are supplied (Eco-Systems 1982).

Inspection of Table 4.2 shows that those techniques already practised by local farmers, which the colonial government also encouraged and wished to extend, were accepted. Those introduced by government which were similar to those already employed (e.g. contour ridging) were also accepted. Others ran into practical difficulties. For example, contour banks in the Mbeya region led to rat infestation; wide grass strips and contour hedges often used up too much land. Boxed and broad-based terraces required an enormous amount of labour, reduced yields and proved unstable. Coercion had to be used to get local farmers to undertake these (technically dubious) measures.

Many conservation schemes in southern and central Africa demanded a reduction in stocking rates and thus in the major form of storage of wealth, capital and prestige (Stocking 1981a, for Swaziland; Cliffe 1964, in Tanzania). In Java, Pickering (1979) reports in detail problems of the same kind. Ridge and tree planting on very steep slopes met resistance when the trees grew enough to shade food crops planted in between and were savagely pruned back. Constant supervision of seedling trees was also a problem. However, this technique was better accepted on poorer lands where food crops were less attractive relative to fuelwood. Bench terraces were technically effective, but ran into problems with the marginal farmer who was unable to sustain the temporary loss of food crops which terracing usually entails. Lastly, a scheme for silvipasture involving quite an intricate technology combining livestock-rearing, fodder grasses and fuelwood, needed considerable capital and physical back-up on the part of farmers, the larger operators of whom were successful. Bonsu (1981) reports the results of experiments to reduce erosion in a savannah soil of northern Ghana by means of a number of agronomic practices, including the leaving of crop residues as a mulch. One wonders how successful this was bearing in mind that residues are used there to feed livestock, as fuel to cook food, and to construct roofs for houses. Indeed, one could posit the hypothesis that this technique, if enforced, would put more pressure on sparse thorn/acacia resources, or possibly animal dung as cooking fuel

Table 4.2   The adoption of soil conservation measures in colonial Tanganyika (Berry & Townshend 1973)

Measure	Degree of adoption	Local/ Government	Description
Contour hedging	+ (if hedge was narrow)	Local (Musoma)	Planting trees, thicket, bush and/or tall grass along contour to stop downslope soil movement.
Contour	s	Local (Tuta)	Building up of top soil into ridges along contour, crops planted on ridges which inhibit water movement downslope and provide plants on ridges with adequate moisture.
Pit system	+	Local (Wametengo)	Cut grass laid in a grid of 7-10 ft squares. Centre of squares dug out and soil piled up on grass lines. Crops planted on lines. Water and washed-off soil collects in pit. After harvest, pit filled in and process repeated.
Tied ridging	+	Local (Wakara)	Tying of contour ridges by building other ridges at right angles to produce 'lattice effect'. Crops planted on the ridges. Similar to pit system.
Trash bunding	+		Lines of trash laid along contour and plants grown on the upslope side. Soil was collected in these lines of trash.
Contour banks	s?	Local	When enough soil had been deposited on trash bunds, or by the farmer removing soil, a bank was formed for crops.

Narrow-based terraces	s+	Local (Erok)	In steep areas, if more soil was washed downslope, small terraces were formed. Local, but also encouraged by government.
Broad-based and bench terraces	–	Government	Larger terraces involving digging soil out hillside. Disturbed soil profile, and also topsoil (used to construct terrace) was often washed away.
Boxed terraces	–	Government	Similar to tie-ridging but involving difficult construction work.
Gully stopping, and Check dams	+	Local (Wakara)	Many forms. Brushwood, with or without stones pushed down into gully floor to trap sediment, which then can be further stabilised with grass.
Diversion ditches, and stormwater drains.	+	Local	Construction of ditches on slopes which were contour banked or terraced, or alongside roads to control runoff.

+ = accepted by farmers; s = measures said to be successful by colonial administration; – = very unpopular with farmers

rather than as a fertiliser. The commonly promoted answer to these problems is more research into the decisions and practices of users (UN Desertification Conference, Nairobi 1977). This is perhaps one of the few areas in which more research may provide genuinely useful insights for more realistic conservation programmes.

(c) Conservation is hampered by existing land tenure conditions. In many countries with very unequal land holdings, land reform is considered an essential prerequisite to a successful soil conservation programme (e.g. Baker 1980, for Peru; and Eckholm 1976: 55), in order to encourage secure and long leases of land so that tenants would invest in conservation and in some cases provide marginal farmers with more land and therefore income to be able to afford conservation measures (FAO 1977a: 27).

Eckholm (1976: 55) recounts the case in Ethiopia before the land reform of 1975 when peasants, forced to work on a reafforestation project, planted the seedlings upside-down as a protest against oppressive land tenure conditions. This example shows the indirect effects of tenure upon the attitude of subordinated peasants who had to plant trees on someone else's land. The same author goes on to say that there are cases where land reform, or at least the break-up of large latifundia or estates, had actually accelerated environmental decline, as in Bolivia in the 1950s or after the emancipation of serfs in Russia in 1816. The major constraints which tenure conditions impose upon conservation programmes that discourage private initiatives to invest in conservation practices are short leases which tempt the tenant to remove as much plant material and nutrients as possible from the soil within the period of tenancy; insecurity of tenants in longer leases; uncertainty of compensation to tenants by landlords for conservation works at the end of the lease; and a lack of an accepted means by which landlord and tenant can share these outlays in the first place (Barlowe 1958).

(d) A lack of participation by land-users in government sponsored conservation. This explanation clearly overlaps the previous one where conservation cuts across existing agricultural and pastoral practices. Presumably preliminary consultation by conservation agencies and institutional involvement of local people in plan formulation and implementation would avoid these clashes. However, it goes further than that and implies a lack of awareness of the need for conservation, of future private benefits for individual households and the village (or larger local community) as a whole, leading to a failure to mobilise people to give their labour for construction (even when it will be paid for), for maintenance work, and for enforcement where discipline is required (to control lopping of small seedlings, to exclude livestock from reafforested and other areas and so on).

> Bringing about community participation in forestry programmes would usually require a more equal pattern of land ownership and control than exists in most LDC's, that is, require redistributive land reform. Cooperation is rarely found to succeed amongst those who are unequal in material terms since it becomes difficult both to ensure an equal distribution of costs and benefits. (Agarawal 1980: 161)

This is an important but different explanation from that which attributes failure of conservation programmes to the conservation agency (e.g. the Ministry of Forestry, the Environment, or *ad hoc* project staff) being ineffective in getting local people to cooperate *with them*. Indeed, internal dissensions between different groups of local people may lead to indifference on their part to government personnel and their efforts, but the causes and remedies of the two explanations are different. In Nepal, a number of inequalities in the distribution of costs and benefits of various conservation programmes and related projects for increasing agricultural and fuelwood production can be identified (Blaikie 1981, 1983). Households with less privately-owned land which had to rely on the 'public economy' (the forest) for livestock rearing rather than private land, were less willing to see restrictions placed upon access to forests or to others' private land for stubble grazing in cases where fodder tree planting on terrace backs was being advocated. According to one source, it was an immensely slow and difficult process for the British and Nepalese project personnel to get the village *panchayat* (village council) to agree to communal efforts necessary for conservation, seed replication programmes, seedling nurseries, terrace improvements, terrace-back planting, woodfuel lot planting and stock improvement, because the vissiparious effects of perceived (and real) different costs and benefits of various parts of the scheme. This was used as one of the major defences of the British aid programme in Nepal which was to concentrate their efforts in villages with a predominant presence of ex-Gurkha soldiers which would encourage both better rapport between British field officers and farmers, as well as an (ex-) army-style sense of organisation, discipline and work-practice within the village itself. In these villages, it was argued, a lack of common purpose would be less pronounced.

Taking a counter-example, the success of a project reported by Veblen (1978) for an area in western Guatemala was attributed to the institution of communal forest holdings, or *ejido*, which has favoured preservation of the forests of Totonicapan. Communal over-exploitation was averted first because of inter-personal face-to-face relationships involved in the design and implementation of forest preservation by a small community where enforcement and the meting out of punishments was a matter for the forest users themselves; second, because of the existence of a special interest group (carpenters); and third, the shared and fierce desire of the people to remain as independent of outside influence and control as possible (p. 434). Another isolated success story is reported by Shaxson (1981, reported in Morgan 1981) in India where a village finally decided to restore its hill-grazing area and plant fuelwood trees, the scheme being designed and implemented locally. It is instructive to compare these accounts with another proposed project (FAO/UNDP) reported by Velloso (1981) in Honduras where an effective linkage between the community and Honduran entrepreneurs for the proper exploitation and industrialisation of forests was hoped for. The Comayagua Plan includes a hydroelectric plant to fuel a planned woodpulp and paper factory; an experimental nursery to test possible future forest products; and the creation of employment opportunities to compensate local people for the necessary reduction in their livestock rearing. The achievement of local participation seems very remote, and a lion's share of the benefits will probably accrue to more powerful elements (entrepreneurs and outsiders). Other more coercive measures are almost bound to be used when participation breaks down – a general issue which will be returned to later.

Turning to the relationship between direct land-users and conservation agents, lack of participation of the former has been attributed to governments not involving the people in the design and implementation of the project. Where government introduces measures that exclude people from resources long since used by them, the people come to view the project as a zero-sum game (Thomson in Glantz, ed. 1977), where their loss is exactly others' gain, and conservation will not in the long run lead to an increase in their incomes. A lack of government efforts to involve local people is often contributory to the design of projects which cut across existing agricultural practices and tenurial arrangements. But also in another sense, conservation has frequently been seen as an imposition, so much so that many local people rioted, formed armed resistance and used the issue as a nucleus for organising wide-ranging political dissent in eastern Africa against the British colonial administration (Young and Fosbrooke 1960; Cliffe 1964). In the colonial context, the word 'participation' had an uncomfortable ring about it, quite alien to the ideology of the political 'status quo', and the colonial administration and agricultural extension service knowing what was good for the African cultivator. Another example, in Algeria shortly after Independence, is discussed by Stewart (1975: 31):

> The central criticism of the 1960 programme has always been directed against its allegedly authoritarian character. The treatment was to be decided by experts, and the people had simply to accept it. Their acquiescence was to be encouraged by a campaign of information and by mobilising convinced peasants to persuade recalcitrant ones, but nothing could stop the tractors from moving in on the appointed day.

Successful attempts to get local people to participate in official conservation programmes may have increased in post-colonial times, but there are many cases of governments choosing coercion rather than participation, for example:

> The soil conservation efforts (in Lesotho) were no more successful in increasing agricultural output than were those installed by the British during the pre-independence period. And again, since no erosion research data were collected, their effects on soil losses are unknown. But the record shows clearly that they were definitely more expensive and at least as disruptive of local village life...as the earlier British programs had been. (Nobe and Seckler 1979: 157)

The case studies by Rapp, Berry and Temple (1973) which were referred to earlier in this chapter also illustrate all the major reasons for failure of conservation in Tanganyika and post-colonial Tanzania

– technical incompetence and incompatability with existing agricultural practice has already been mentioned – also attempts at coercion of peasants to build erosion works (p. 117, 250), and a lack of participation, with an excess of petty restrictions (p. 251).

Even where the government pays for the mechanical means of conservation entirely (and thus avoids the problems of a lack of capital or labour on the part of the farmer, his inability to forgo food crops for a season, and the reduction of his perception of the risks involved even if his private resources are sufficient), such an act engenders apathy on the part of the farmer, who regards the work as belonging to the state (FAO 1966: 172, for a project in Morocco).

(e) Institutional weaknesses. These causes of failure are attributed to 'problems at the top', although of course these must be linked to problems further down, particularly of lack of participation of local people, inappropriate conservation measures, local enforcement and so on. It perhaps seems the easiest problem to tackle if the viewer *also* is located somewhere near the top (e.g. a foreign advisor, international academic, or senior bureaucrat or politician in the country involved). This perhaps explains the predominance of recommendations for institutional reform, compared with the other deeper-seated changes in civil society which are implied (and later analysed) in this book.

A commonly recognised problem is a lack of coordination in conservation policy which comes about because the wide-ranging issues implied in its brief are traditionally dealt with by separate 'line' ministries. Here soil conservation does not stand alone in this, and rural development in general suffers from the same problems which have generated a large literature in the improvement of planning and administrative systems. In some cases, there are competing institutions for policy-making and implementation of soil conservation (such as in FAO itself) as in many governments of lesser developed countries. In response to this a separate central organisation with special powers is often suggested. Sandbach (1980) cites a United Nations Environment Programme (UNEP) report published in 1976 which states that seventy countries had created central organisations concerned with the environment. For example, at the time of writing every nation in South America except Paraguay has such an institution. A different but not exclusive approach is to insist upon a conservation element being included in most rural development projects. This is sometimes quite difficult if funded by governments, but even more so if funded by bilateral aid, where the donor may have its own ideas on the subject. Furthermore they may be different ones from other bilateral or multilateral agencies, making a coordinated national conservation programme rather difficult. The case of Nepal illustrates this. In 1982 there were sixteen forestry/conservation projects and conservation elements in six integrated rural development projects (each with a different donor agency). Some of these are included in the Sixth Five Year Plan, and some of which have been tacked on to *ad hoc* annual plans. There are scarcely monitored nor integrated into a national conservation plan. In Kenya a new Ministry of the Environment and other far-reaching institutional reforms are currently before parliament. Inter-departmental squabbles are blamed for failures in India (Bali 1974; Ranganathan 1978).

Effective legislation is another area of institutional strengthening often advocated. Although there are many laws on the statute book (often dating from colonial administration), they have either become dead letters or are not integrated into a coherent national conservation programme (FAO Soils Bulletin No. 30, 1976b, lists various legislation for different countries). Indeed, many of the conservation statutes in Kenya, Zambia, Zimbabwe and Tanzania, which the post-colonial governments have resurrected or only slightly redrafted, date from the British administration. These statutes were ironically among the main foci of opposition during those countries' independence movements. In the case of Thailand, legislation has been on the statute books for at least twenty years, but its Constitution of October 1976 and Interim Constitution of November 1977 contain no mention of environmental protection at all, as also is the case of the Philippines (Shane, in Mac Andrews and Chia Lin Sien, eds. 1979). However, from evidence cited by Shane (ibid), it is not so much the lack of legislation but the unwillingness or inability of governments to enforce existing measures adequately which is the more fundamental problem.

5. Policy implications: from black boxes to systems of black boxes

The response to these five perceived difficulties has been various. At the most analytically simple, the difficulties remain as unexplored discrete entities or black boxes, which then have to be removed or steered round in the formulation of a conservation project or policy. In rather the same way as in some Basic Needs Strategies, a checklist of desirable attributes of a good strategy can be compiled (Blaikie, Cameron & Seddon 1979). A drawback to this response is that merely bearing a desirable attribute in mind does not necessarily remove the problem. Its advantage is that it is at least problem-orientated and gives the impression (which is sometimes justified) that one has the means to solve it. The more abstract the mode of analysis of the problem and therefore the more connected the items on the check-list become, the less likely that one can solve them alone. However, even a rudimentary enquiry into the problems discussed above shows that they are connected in obvious ways. Local participation is achieved because, and as a result of, appropriate conservation measures which do not cut across existing practices and irreducible means of earning a livelihood. Effective programme formulation and implementation by government to provide external inputs to appropriate skills is essential for local participation and implies effective and coordinated administration, and so on.

Furthermore, all the items on the check-list themselves, such as local participation, effective administration by government and consonance between conservation measures and existing agricultural and pastoral practice, require deep-seated political economic preconditions. Community participation in conservation can occur under a variety of different circumstances, but one can identify some that make it easier. For example, inequalities of political power and access to land should not substantially weight the costs and benefits of conservation measures to or from particular groups or classes at the local level. Also there should be an effective and widely respected authority which can solve disputes and give penalties in the case of infringement of mutually agreed codes of practice. Finally there should be a trusting two-way communication between externally provided skills (e.g. water engineers, agronomists, planners,) and local land-users. These three facilitating conditions are predicated upon the nature of the local political economy–who has power and how it is exercised; upon the nature of the state, and the interests and ideology of people who serve in government (e.g. bureaucrats, forest police, or politicians). Hence there is a painful dilemma for the policy-maker between treating a problem as a black box which will not go away but whose size does not seem intimidating and is amenable to exorcism by benign rhetoric and minor policy adjustments; or as a more diffuse, analytically complex and sophisticated *explanation* of why most of the problems exist. The choice that this book tends towards is the second, and the ensuing problems are discussed in Chapter 5.

The discussion so far does not imply that piecemeal improvements of conservation programmes have not occurred. While research on the agronomic side of conservation has taken considerable strides during the past fifteen years for all its problems of misapplication, there have also been some advances in the thinking on the socio-economic side of soil erosion and the resulting design of conservation programmes as well. This is particularly important in the ideological vacuum which was left by the retreat of empire from Africa and Asia where much soil conservation had been designed by colonial administration. Three different ways can be identified in

for success

which piecemeal improvements to the conceptualisation of the soil erosion problem have been transcended. They are: (a) a new attempt to understand the interrelatedness of the current problems that beset conservation policies; (b) to stimulate flows of information between land-users and institutions of government; and (c) linking conservation with wider development efforts.

An awareness of the way in which the reasons for erosion are connected has implied the necessity for inter-disciplinary study. This has led to a change in the way in which both physical and social data are collected and combined in an effective conservation policy, or, as Pickering (1979) puts it, in such a way as to ensure the technical validity of a conservation technique is appropriate over the same area as its social validity. One approach is to hire a sociologist to peer into the black box (the quotation from Heusch 1981: 423, and p. 46 in this book about the role of the sociologist is apposite here), in a similar way to anthropologists who sought to tell colonial administrators how to avoid irritating the people they governed (although, according to Feuchtwang 1973, the administrators did not take much notice of what they had to say). It is the experience of other sociologists on similar missions that their findings are frequently cast in a negative role – that certain conservation measures will not succeed or will actually do harm – and also that they are frequently overridden or relegated to writing a disparate chapter in a project document or report entitled 'Social constraints to soil and water conservation'. Under these circumstances at least a gesture towards listening to the local people has been made. In other reports, however, these findings are actually incorporated into a project or programme – although whether those parts of the project are actually implemented is another matter.

Another approach to handling inter-disciplinarity is to educate agricultural engineers and administrators into the way of being sympathetic at least to these social factors. A recent publication spells out the need for more inter-disciplinary awareness (UNESCO 1979a: 5):

> Engineers are often most concerned with the attainments of a high level of technical competence which has resulted in the gradual erosion of the general level of social awareness and perspective necessary for an engineer to serve his community usefully... There are numerous examples where there has been an effort to apply an engineering solution to a problem without taking adequate account of all its parameters (tube-wells and salination, road building and rural income distribution).

Seminars, courses, systems analysis, panel discussions, case studies and simulations are some of the solutions suggested. How far this approach can hope to have an impact is rather doubtful. Also engineers are singled out here as the culprits when it could well have been social (and technically incompetent) scientists!

Lastly, inter-disciplinarity is assisted not only in the logistics of using knowledge, but in the epistemology and methodology of that knowledge itself. Here systems analysis has helped considerably. Such works as Odum (1971), Dumsday (1971), Dumsday and Flynn (1977) and Rieger (1978-9a & b) are examples in that field. Systems analysis has important advantages in that it can handle both complex biophysical relationships (which is more difficult in a rigid mathematical programming approach) and can include the effects of policy instruments and other hitherto exogenous social and economic variables. The disadvantages are that the technique is very ambitious, is sensitive to data inaccuracies and incompleteness, and technical problems of model validation are formidable. However, even if it is a matter of putting different variables in boxes on a sheet

Fig. 4.2 A systems approach to soil erosion in the Himalayan/Gangetic region (Rieger 1978/9)

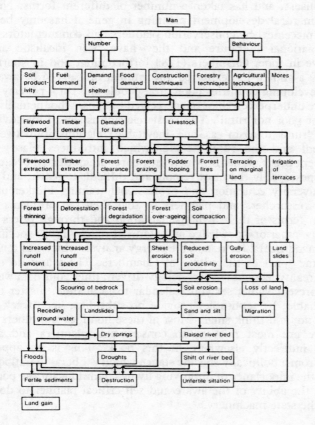

of paper and following through likely relationships qualitatively – a far cry from what is usually meant by systems analysis – it is a further intellectual advance on the black box approach to 'social factors'. Fig. 4.2 from Rieger (1978/9a) is an example of this approach. It gives a good intellectual overview of the problem and a 'map' in which can be located detailed close-up studies, which at the same time do not lose sight of the general context in which they are made. In the context of lesser developed countries with acute problems of data collection and interpretation, a systems approach rather than systems analysis will be more viable. Modelling farming decisions by means of farming systems research is also another promising avenue. In this approach farming decisions are not analysed in isolation from each other or from non-farm employment (something which is advocated in Ch. 5 especially, and operational-ised in a decision-making model in Ch. 6).

The second trend in current conservation thinking concerns the relationship between the programme and its clients or government and land-users, and has taken a number of different forms. New thinking in rural development planning in general has only been taken up piecemeal by conservation planners and commentators in the international literature and they have been idealistic and exhortative in tone. Chambers (1983 forthcoming, and in shorter form 1980 a and b) has exhorted government officials to be aware that 'in developing countries poor rural people and rural poverty in general are underperceived or misperceived by those who themselves are neither poor nor rural'. The most relevant aspect of his advice concerns 'rural tourism' as one inevitable relationship between government and the rural people, which introduces biases in development programmes in which the poorer are neglected because they are inescapably the most remote and difficult to reach; and in which university education and professional training blinker and condition researchers and planners so that they see problems of poverty in terms of their own specialisation and through filters of supposedly superior knowledge and status. French (1982) specifies 'ten commandments' for renewable energy analysis which has much of the same spirit as the advice of Chambers. These are: do not harm; be reality led and not technology-driven; filter your data well; find net present value; set your calendar to local time; master the unquantifiable; be replicable; provide no subsidies; keep track of what you do; and think small. Most of these homilies and others of this genre are most relevant to conservation planners and are discussed indirectly elsewhere in this book. This is an appeal deriving from a politically specific stance which is liberal in outlook, implying the importance of improving living conditions for the poor, and from the ability of the aware and self-critical planner to do it through the state machinery.

Other developments of better government outreach to rural land-users advocate the need for more *local* research, although the methodology of local research remains unspecified. Chapters 6 and 7 attempt to develop one which enables local situations to be studied in relation to wider concerns, and, to use ugly but accurate jargon, to 'theorise the conjuncture'. Such calls have been made by Hudson (1981) for conservation, although they are far more frequently made by agricultural economists and agronomists who are concerned with interactive research by farmers and research stations (Biggs 1981). A 'top-down' comment by Holdgate (1982) also makes a plea for relevant local research which can be applied and its resulting policies replicated.

Lastly, conservation has been seen as just one element in rural development efforts, and conservation elements have been linked with agricultural extension, institutional development (cooperatives, regulated markets, government-supplied services, and so on), and the provision of infrastructure (storage, roads, buildings, etc.) (Adamson *et al.* 1982). This has one most significant aspect in that soil conservation is not singled out as a specific and separate problem to be solved by a particular policy – it is conceived of as normal practice and must be incorporated into the business of improving incomes for farmers. Many integrated rural development projects have conservation elements built into them, but, judging by evaluation reports, these are acknowledged to be difficult to measure, slow to bring benefits, and are often neglected or abandoned. Many bilateral and multilateral aid agencies have rightly upgraded the importance of monitoring and evaluation in their activities, and as Chapter 2 has indicated, quantification and measurement of the impact of soil conservation are essential. None the less diffuse benefits from conservation accruing over a long-time horizon and targetted very widely among the rural population are poor candites for adoption and funding. What cannot be justified quantitatively cannot command large resources. Thus the inclusion of conservation elements in rural development projects tends to be more a conceptual advance than a practical one, and to be an item in a project document rather than an output.

All these three new developments are in the direction which this book advocates, but they do not go nearly far enough. A further extension of these new developments requires a new approach to the whole issue of erosion and conservation. This approach must first identify and question three fundamental assumptions on which most conservation thinking currently rests. These are:

(a) The causes of the problem of soil erosion lie only in the actual place in which the physical symptoms are felt. A new approach must also recognise they also derive from wider non-location-specific

political-economic relations, and also in other places that may be far
from the affected area.

(b) Soil conservation programmes and policies, particularly those
formulated by expatriates, assume the state to be neutral, able to
mediate between competing interests, and to intervene above them
to apply rational policies for the maximum aggregate benefit. Thus
conservation problems can be more readily solved by better planning
carried out by the apparatus of state (ministries of agriculture, local
development offices, forest rangers, or research stations).

(c) Putting (a) and (b) together, another related assumption
becomes clear. Land-using practices are economic activities, in
which political-economic relations may be crucial (e.g. the ownership of the land, the distribution of that ownership, the crops
grown for the market or for self-provisioning, the existence of wage
labour, and so on). Conservation policies *must* affect these relations.
Some of those involved lose, some win, others will be unaffected.
Economic and political interests are therefore affected – in the
countryside and in the government. By not considering these
interests explicitly, most thinking on conservation policy cannot link
conservation with a whole set of wider issues deeply embedded in the
patterns of power within a state. In the next chapter these
assumptions are further questioned and new ones are suggested.
Upon these a new approach is constructed.

Chapter 5

social + environ...

A new approach—with new problems

what is this exactly?

1. Two paradigms

One of the major problems of building a theory of soil erosion is the high degree of contingency which always accompanies any explanation of soil erosion at a particular place. As Chapter 2 has indicated, soil erosion has occurred in many different societies and periods of history – in feudal Europe, the Soviet Union, Tsarist Russia, the United States in the 1930s, and in present-day peasantries and under widely differing physical conditions. Essentially, in an explanation of a case of soil erosion, there are two sets of specificity to be tackled – that of the physical system and that of the social/economic system, and they both have to be brought together and analytically integrated. An omission of the first would lead to a failure to specify the physical processes of soil erosion, their spatial variability and interaction, and indeed the immediate causal variables such as slope, soil structure, vegetative cover, land use, rainfall intensity and so on. To take an extreme illustrative case, one can imagine a situation where broadly similar relationships exist between direct land-users (pastoralists or peasants) and other groups and the state, and they develop historically in a steep-sloped area and in a plains area nearby. In the steep-sloped area, a situation might have developed where overgrazing and deforestation occurred as a result of onerous taxation and loss of land to colonisers, leading to virtual environmental collapse. The same economic and social conditions in the plains area might have led only to loss of soil fertility but not to wholesale soil erosion and environmental collapse. Therefore the subsequent histories of the two areas will diverge and a different relation between environment and people will develop in each of the two areas. In this illustration a very obvious physiographic variable cannot be ignored and is integral as part of the explanation, as well as an explanation for human action. The omission of the second (the social/economic system) leads to a purely technocratic and physical study of the processes of soil erosion and perhaps the immediate land-uses leading to it, without any analysis of other political economic relationships at the local, regional and international scales which determine the actions of the land-user in the affected area.

The two elements tend to lie in different scientific paradigms. The first is typically within the purview of the natural sciences and is the subject matter of agricultural engineers and soil scientists, while the second is within the social sciences and is studied by economists and sociologists. There are considerable epistemological and ideological problems involved in the combination of these two elements, and these are discussed later.

There is also another separation of intellectual concerns which must be brought together in an integrated analysis of soil erosion and conservation. There is the concern for the physical manifestation of the problem, a place-based concern for where it occurs. This tends to focus on the physical symptoms of soil erosion and finds its expression in maps of soil loss or erosion hazard, but it also includes other spatial, tangible variables such as land use, stocking densities and the geographical expression of the flow of energy between people and the biosphere. Therefore it is location-specific, place-based and conjunctural.

There are three ways in which the place-based, location-specific element is relevant to any analysis of soil erosion. First, it clearly focuses on where the physical symptoms of the problem are. At smaller geographical scales it is physiographic variables (those describing erodibility and erosivity) which best explain the spatial variations of soil erosion. This is the conventional (and essential) domain of soil erosion studies as presently conceived.

Second, a place-based concern must be directed to the study of *other* places where perhaps there is no visible soil erosion problem, but where there are processes of spatial displacement of land-users. For example, efficiently run coffee estates which conserve soil well, can 'cause' soil erosion in other places, by displacing small peasants. A capital intensive fruit plantation in the valley floor of a central American Republic can 'cause' soil erosion on hillsides which are being cultivated by peasants who have been dispossessed of more suitable land for cultivation. In Lesotho, forestry programmes have taken grazing land out of use causing immediate pressure on remaining land, without a displacement of the actual homes of land-users.

Third, other places which feel the effect of soil erosion downstream in the form of deposition and floods must also be part of the place-based analytical concern for soil erosion for two reasons. First and more obviously, direct curative measures (e.g. flood protection) clearly are within the domain of a soil conservation policy. Second, there may be various political reasons for including these areas in a comprehensive analysis for accounting purposes. For example, the benefits of a soil conservation programme will be felt not only by the direct land users who are causing the soil erosion, but by others who may be subjected to less severe crop losses, deposition of gravel on cultivated land, loss of livestock and building or siltation of canal irrigation networks. Therefore political and economic accounting of the costs and benefits of soil conservation programmes for others outside the directly affected areas is potentially relevant. Sometimes those persons may live in a different country (as in the case of Nepal and India, for example), which may well create both problems and opportunities for the funding and joint cooperation of soil conservation ventures. An estimated $US600m. annual damage occurs in India due to flooding and siltation, much of which is caused in Himalayan watersheds in Nepal (Gribbin 1982, Thompson and Warburton 1982).

There is also the concern of the non-place-based or non-location specific networks of economic social and political relations acting directly and indirectly upon land users. It concerns relations between people, and not just between the direct land users and the environment. Land-users are taxed, sell their produce in the market, work for or employ others, have unequal access to land or other agricultural inputs and are part of the processes of agrarian change such as commercialisation, accumulation, disinvestment and differentiation. Many of these relations directly or indirectly affect land using decisions which lead to environmental degradation. It is these processes which are amenable to the analysis of political economy, and involve the objective identification of different economic interests in the countryside as well as the town.

It is true that these processes may well have a spatial expression in a specific situation, but there are others to which it is difficult to attribute any physical manifestation, or by definition which cannot have a spatial expression at all. For example, the state may regulate the ownership of land through land tenure, or give foreign franchises and concessions, and may pursue national pricing policies for agricultural produce. These processes, too, are subject to political and economic analyses and are often intimately bound up with land-users who are responsible for accelerated soil erosion. Both these elements are needed in an understanding of soil erosion conservation, but hitherto it is only the first which has usually been considered – and only in the first of the three location-specific senses at that.

2. The integration of social factors

These 'social factors' then, are part of both the place-based and non-place-based concerns described above. In the former, they find expression in land-use patterns, the spatial patterns of agricultural technology including the diffusion of innovations, price-distance relationships of inputs and outputs, spatial patterns of size of landholdings and other more complex 'eco-class' relationships (involving spatial displacement and marginalisation of weaker groups, which will be discussed later). In the latter, they *are* the political-economic structures within which land-use decisions are made. A check-list of these social factors is given below, but any analysis of the processes mentioned above will combine the items of the check-list.

Following a 'bottom-up' analytical approach, we start with the smallest decision-making unit which uses land, usually centred on a hearth, involving a family, extended family or occasionally a larger unit. This unit is the smallest one which collectively makes decisions about its allocation of labour and other privately controlled resources (though it may not necessarily own them, in the sense implied by private property). There will be definitional and ultimately theoretical problems in a few cases where land-use decisions are shared in a complex manner between a small group like a household, and a larger community. However, in the majority of cases, we are talking about a household constituting a nuclear or extended family, or a pastoral group controlling a flock or herd.

Although land-use decisions are made by the household, these may well not be made equally by all members of it, and this may be significant in the perception of conservation and any government sponsored conservation programme. So, a prior starting point should be the politics within the household itself. The general division of labour in peasant and pastoral households is significant – it is women who frequently collect water, fuel, forest litter and fodder, and indeed in some societies do most of the agricultural work as well, except ploughing (as, for example, in parts of South Asia). However, cropping decisions, the purchase of inputs and marketing are often done by men (with important exceptions, for example in West Africa). In cases where women (and the children they directly control) collect forest products of one type or another, it is they who are most aware of any deterioration in resource availability, but who are often an inaccessible but important target group in agricultural extension efforts to get conservation methods accepted. There are other possible inequalities within the household, frequently concerning both gender and age. In many societies it is the older males, including the 'head of household' who allocate labour, control access of young men to women for marriage, and make other major decisions. It is their voice which is heard, and they often are the opinion-makers in the household and in the local community. Their views frequently are less innovative (not necessarily in a pejorative sense) than the next generation over whom they have considerable *assess politics.*

control. Whatever the politics of decision-making within the household, male dominated, elder-dominated, or more egalitarian, it is usually necessary to pursue these enquiries first.

At the level of the household, land-use decisions are often only a sub-sector in a range of income opportunities, and this range has to be fully specified to include the most common forms of income generation since land-use decisions are affected by these other forms. Thus non-agricultural activities should be considered alongside those which directly use land, and this is particularly important where there is widespread rural-urban migration with or without a remittance economy (e.g. South Yemen where outmigration to Saudi Arabia and the Gulf has led to the abandonment of terracing; the Rif in Morocco and Algeria; the Nepalese hills; Lesotho and other labour reserves of Southern Africa; or the Andean highlands where many of the male population have migrated seasonally or permanently to the coastal cities). Chapter 6 formalises a decision-making model in which a household may take up a number of different income opportunities according to its capability and the 'entry costs' of such opportunities. Many of these opportunities are expressed in terms of different crops and associated agricultural practices (inter-cropping, mulching, or tie-ridging) which have direct implications for soil erosion.

The political-economic structures *behind* the range of choices is the next necessary analytical step and opens up wider concerns at the local, national and international level. At the local level, certain decisions concerning land use and deployment of labour may be made by village councils, whose range of authority, extent of public acknowledgement of penalties, acceptance by people, and their distribution of power have to be considered. Sometimes effective land-use management and watershed management requires a number of institutions of this kind to agree, which may well be even more difficult because of unequal costs and benefits of conservation measures between different communities or villages.

3. The state, government and administration

The nature and interrelationship between these entities is a difficult and debated subject, and readers are referred to some of the best-known texts on the subject. Here a simple conceptualisation as is serviceable for the purpose of this book is made and it follows the main elements of Miliband's [1969]. However, it cannot be proved or disproved in any final way and is unavoidably a political statement. None the less, it is necessary to make explicit what the state, government and administration are, and what they do. Many commentaries without such an explicit and analytical statement, bewail the lack of political will, expertise or rational planning, without offering any explanation as to why these qualities frequently do not exist to carry out effective conservation.

First of all the definition of some terms is needed. The state is a most problematic term since, as Miliband [1969: 46] put it, 'the state is not a thing (and) does not, as such exist'. The state stands for, in an abstract sense, the final repository of agreement of the people to be ruled, and therefore an ultimate legitimation, backed up in the last instance, by the state's claim to the monopoly of legitimate use of force. So for example, the inclusion of conservation principles in the constitution is a symbolic statement to which all people 'should' give their allegiance, and it is the duty of all governments to uphold and abide by it.

The government on the other hand speaks and acts in the name of the state. It may be a weak government and, although formally invested with state power, does not or cannot use it. The government or political executive has as its instrument for the business of governing, the administration or bureaucracy, which extends through the ministries and departments, to a variety of institutions such as public corporations, banks, and others (particularly the institutions of education and the media). The crucial difference between a social democratic view of the state and the one taken here, is that we recognise that the bureaucracy itself is political in the course of exercising its executive powers, particularly in the realm of policy-making at the upper levels and in implementation at the lower levels. By extension, then, conservation practices left undone, legislation remaining unheeded, projects that only serve to keep research officers in salary and which never leave the experimental station (those things *not* done) are also political acts and not just omissions, or non-events which do not need explaining.

Although there still exists the notion that administration should be or is apolitical and merely a neutral tool of the government, this is palpably not so in practically all states, industrialised or less developed or 'socialist'. Senior civil servants have tremendous power, particularly where government is weak, changes are frequent, or their political leaders lack technical or administrative expertise (or have misplaced confidence!). The administration itself is of course no monolith and does not act as a single bloc, pursuing its interests whatever they may be. At the top there are very senior bureaucrats who, with their counterparts in the military and judiciary, form a state elite. In some lesser developed countries, particularly in central and east Africa, it has been called a 'bureaucratic bourgeoisie', a term which is discussed below. Recruitment to its ranks is highly selective in terms of social and educational status, and it perpetuates itself by spending parts of its accumulated surpluses on privileged education, often abroad, for its favoured sons (and occasionally daughters). The term 'bureaucratic bourgeoisie' to describe this state elite has certain problems, and used strictly should refer to managers of state-owned enterprises. However, this class may actually enter into economic activities for private gain using, for example, its privileged position to secure credit at favourable rates of interest, obtain monopolies and franchises in import and export businesses, start up industrial enterprises, speculate in urban property and so on (for a debate about the bureaucracy in lesser developed countries, see Coulson 1975; Leys 1975: 193f.). An example comes from Williams (1981, p.23) when he discusses the beneficiaries of a large World Bank project in Nigeria: 'these rich beneficiaries are drawn from army officers, government officials, contractors, merchants and members of the office-holding aristocracy, who purchase land in anticipation of benefits from the project and from cheap bank credits'. In many countries this elite acts as the chief negotiator with transnational companies and therefore its members become beneficiaries from negotiations with logging contractors, purchase of agricultural estates, permission to sell agricultural inputs, purchase of agricultural inputs, and other policy measures.

The distinction between individuals in government and the senior civil service, banks and corporations is often more blurred than in western countries, and therefore this elite tends to form a coherent and powerful set of economic interests, (although these may be contradictory and are usually very diverse) and are not only represented in government but often *are* the government. At the same time this is not to deny there are administrators and administrations which claim to act, even want to act, in a disinterested way. An analysis of Indian administrators (Wood 1976) shows how the liberal democratic ideal of the disinterested administrator carrying out the orders of his political masters, left him, in independent India, with relatively low-status and low-income compared to those in commerce or business. This created a tendency for the administrator to facilitate the flow of benefits which the state frequently confers upon private enterprise to private businessmen for a return, and, indeed, involve himself in a variety of legal and illegal activities for which the post of senior government servant is advantageous.

Lastly, after government, the administration, military and judiciary, a fifth element of the system can be identified which Miliband (1969) calls the various units of sub-government – in a sense the extension of central government but also a voice of the periphery, and thus a channel of communication between the two. This element is most readily identified in a federal structure such as the USA, Brazil, India or Nigeria, but is also significant in local assemblies, as well as in centres of district and regional administration. The politicians and bureaucrats at this level can have a profound effect upon the local people, and can obstruct, alter or intensify central government actions. So the bureaucracy at this level will not necessarily fulfil the economic interests of its masters in any simple way.

Only a small fraction of the bureaucracy can directly benefit from its position of high office and crucial monopoly or oligopoly in negotiations and policy-making. All have career structures, which have varying degrees of opportunity. For those at the lower and middle grades, job security and risk-minimisation in decision-making may be their most important considerations in the execution of their duties, since government employment in many lesser developed countries is relatively secure (compared with other jobs available), with special privileges such as pensions, health services, government schooling and housing. At the lowest levels of the government machinery (the village clerk, the forest policeman, and the agricultural extension agent), there are additional contradictions. For these people day-to-day life is much more concerned with general social intercourse with farmers, pastoralists, small businessmen and local politicians, chiefs, headmen and so on. As a result lower level or petty bureaucrats frequently have to rely upon local people for economic perks (access to a small plot of land, or to credit on favourable terms, or the rent of a house). Local conservation officers, forest guards and rangers therefore find that 'rational' conservation policy is tempered by a web of social and economic relationships with various members of the public at the local level. Thus considerations of job security, avoidance of reprimand, and conditions of work, are further tempered by reciprocity and reliance upon local people. These various forms of differentiation within the bureaucracy are vital in the understanding of the formulation and implementation of conservation programmes.

Conservation programmes seldom bring opportunities for personal advancement for senior bureaucrats, unlike other activities such as the issue of import-export licences, the handling of foreign business, and the control of access to state-controlled markets. Therefore in a positive sense, conservation does not attract bureaucrats or quasi-official middlemen. Indeed many mediocre and/or failed bureaucrats tend to be 'promoted' to conservation by their more thrusting and successful colleagues. However conservation policies also affect the careers of bureaucrats and elements of sub-government in a negative sense, putting them in difficult positions since these programmes usually include policy measures which are directly injurious to the interests of the state elite in their private economic capacities. Usually conservation programmes and individual projects have a number of elements, some contrary to the interests of the state elite, others neutral, and occasionally others which could provide opportunities (e.g. contracts for earthworks, labour recruitment for building of dams, or tax advantages for conservation undertaken on private land). Also conservation programmes involve potential clashes with landowners, politicians whose constituencies are adversely affected, forest contractors who would like to fell trees without official controls and so on. Lastly conservation programmes often fail, and senior bureaucrats may have to take the blame.

Having provided the briefest outline of the relationship between state, government and state machinery, let us show how it is helpful in the analysis of soil erosion and conservation. First, it provides a more convincing explanation of the failure of conservation than 'a lack of political will' or 'powerful vested interests' or other dark and unspecified reasons. There are clear reasons why state power should encompass the means to conserve soil, reafforest and so on, but why do governments not do it successfully? Seeming contradictions can be explained in situations where conservation policies may be drawn up in some areas (perhaps financed by foreign aid, and attached to prestigious foreign fellowships and enhanced salaries for senior officials) – but in a neighbouring area a *carte-blanche* logging contract is offered to a transnational company. Successful conservation may well imply a cutback in commercial crops and therefore a reduction in foreign exchange on which the state elite depends for its imported luxuries, foreign travel and education. Where there is state intervention in food prices, these tend to be low to favour a political alliance of various classes who live in towns. Pricing may well be a critical factor in returns to farmers and to decision-making about extension of cultivation, and precise agronomic practices with direct implications for soil erosion, and yet prices of foodstuffs for urban-based elites are naturally a highly sensitive political issue. Thus what is done, and what more commonly and significantly remains as rhetoric, can be explained.

Second, this type of analysis shifts primary attention away from the preoccupation of conservation policy-makers, academics and consultants with institution-building, training government officers in environmental awareness, tightening up and rationalisation of administrative and legislative procedures. All these reforms may be necessary, but alas, so far from being sufficient that a reassessment of future conservation needs to be made. Empirical justification for this gloomy remark is all too evident in Chapters 2 and 3.

Third, such analysis links national political-economic processes relevant to erosion and conservation to international affairs, mediated in a great variety of ways by the state systems of different countries. Apart from the efforts of transnational companies already mentioned, there are other strategic and financial considerations. As was mentioned in Chapter 3, the American decision to sell grain once more to the Soviet Union in 1977 wiped out the Soil Bank established in the 1970s and caused ploughing up of marginal land and a marked increase in erosion rates (Cook 1983). The American blockade of Cuba threw the latter's post-revolutionary land-use planning into confusion and caused wholesale cultivation of food crops on very vulnerable steep slopes. The deterioration of international relations between the USA and many Latin American countries saw the withdrawal of American aid which financed soil conservation efforts there. Lastly the effects of the world recession and specifically the price rise of oil in 1973 have put pressure upon national funding of conservation programmes; encouraged the use of woodfuel or dung; reduced artificial fertiliser application rates; exacerbated the foreign exchange position of most lesser developed countries and discouraged government and private initiatives in conservation. The advantage of the method of analysis suggested here is that these international influences are not treated as isolated and exogenous instances, as if operated by *deus ex machina*, but are explicable and connected within the context of the workings of the world economic system.

These views are similar to those of Heyer, Roberts and Williams (1981) who discuss rather broader issues of rural development. They also stress that the assertion that rural development serves all or almost all interests is a necessary myth. The contradictions *and* convergences between national and international aid agencies have to be assessed in the design and implementation of rural development projects. International organisations have to deal with 'governments who in most cases do not represent their peoples, and certainly not the poor peasants' (p.2). One of the major reasons for the widespread failure of rural developments is that it usually does not serve the interests of the people at whom it is ostensibly aimed. Instead, it benefits big business, contractors, consultancies, construction firms, government officials, international experts and academics. The point made here and in Heyer, Roberts and Williams (1981) is that these conflicts of interest and how they are expressed in governments (and outside) have to be recognised.

In summary, soil erosion problems can be analysed in a framework of Chinese boxes, each fitting inside the other. The individual within a household, the household itself, the village or local community, the local bureaucracy, the bureaucracy, government and nature of the state, and finally international relations all represent contexts within which actions affecting soil erosion and conservation take place. A specific analysis must identify these contexts and the relationships between them.

4. The expression of class interests in erosion and conservation

It is when the physical phenomenon of soil erosion affects people so that they have to respond and adapt their mode of life that it becomes also a social phenomenon. When this response affects others and brings about a clash of interests – and it usually does except in areas of extreme remoteness and low population density – it becomes a political phenomenon as well. In this section we examine the clash of interests which soil erosion brings about, and we ask whose interests they are and how they are pursued in the face of conflict?

The major determinant of the response to soil erosion is the degree of political power of the class(es) or group(s) involved. Starting with classes and groups with very little power indeed, their response is the weakest. For those, such as the rural semi-proletariat and the absolutely landless with little control over the means of production with which to meet their needs, even individual responses are denied them. In essence their class position often forces them into using the environment destructively and inhibits any adaptive response to its inevitable deterioration. Any individual act of self-denial or investment (e.g. leaving small saplings to regenerate the forest, or investing in an improved method of cooking food) is frequently not repaid – a classic tragedy of the commons does not respect individual and isolated remedial action. Peasants and pastoralists who have more control over the use of private land and/or flocks can carry out both new agricultural practices (mulching, inter-cropping, zero-tillage or improved cultivation), and invest in private erosion works such as terraces, bunds, gully-head protection, etc. However, political action is required to tackle the more systemic and widespread symptoms of environmental deterioration such as the destruction of the forest, wholesale sheet erosion across a large area, or the threatened collapse of a whole tier of paddy terraces. It is then that a conflict of interests may arise. It may arise simply because the risk of deterioration of a resource may be unequally shared among affected people, or because the fruits of personal sacrifices will be enjoyed by others, or discounted future benefits do not justify present sacrifices and are distributed unevenly. For example labour inputs for erosion works and temporary dislocation of food production, fuel supplies and fodder for livestock may be borne by one group more than another.

Very often those without enough political power to influence the course of events resulting from erosion and from official conservation programmes, are those who are politically subordinated in other related ways. For example, in Mali, resources flow from the poor peasant to the urban sector where a variety of powerful classes live and realise that surplus. A state marketing system gives derisory prices for commodities produced by peasants and later exported to earn foreign exchange spent by the bureaucratic bourgeoisie and other privileged classes, or consumed as foodstuffs at cheap prices. However, 80 per cent of the revenue generated is spent on civil service salaries, and yet the government by imprecation if not implication blames this peasantry for sheet and gully erosion. Lipton (1977, 1982) in his thesis of 'urban bias', while being strongly criticised on various grounds, has many examples of this phenomenon. He paraphrases his thesis as follows:

> Small interlocking urban elites – comprising mainly businessmen, politicians, bureaucrats, trade-union leaders and a supporting staff of professionals, academics and intellectuals – can in a modern state substantially control the distribution of resources. In the great majority of developing countries, such urban elites spearheaded the fight against the colonizing power. Partly for this reason urban elites formed, and have since dominated the institutions of independence – government, political parties, law, civil service, trade unions, education, business organisations and many more. But the power of the urban elite, in a modern state, is determined not by its economic role alone, but by its capacity to organise, centralise and control... Rural people, while much more numerous than urban people, are also much more dispersed, poor, inarticulate and unorganised. That does not make them quiescent but it does diffuse their conflicts. (Lipton in Harriss 1982)

Moving now to collective political responses, the problems of political protest by the peasantry have been widely discussed by Shanin (1973), Alavi (1973), Cohen et al., eds (1979). The specific protests against soil conservation policies and other related issues (such as the taking away of land for settlers, plantations or the removal of forests) have tended to follow the same pattern – of violent, politically primitive, and usually short-lived protest involving marches on towns or centres of perceived political power of their oppressors and occasional guerrilla warfare (e.g. the Mau Mau movement in Kenya, although the issues involved were much wider than soil conservation). Examples of peasant or pastoralist protest where official conservation policies were a catalytic or a leading grievance are Kenya (Heyer et al. 1981: 100), and the 1982 attempted coup, Tanganyika (Young and Fosbrooke 1960; Coulson 1981: 56), Zambia (Robinson 1978a: 32); and the Chipko movement against commercial logging companies in the Indian Himalayas (Albert 1979; Lean 1981; and for peasant struggles in general, Cohen *el al.* 1979).

In the vast majority of cases, soil conservation measures were seen by land-users to be a symptom of oppression either by a colonial regime or by small interlocking urban elites. From some of the evidence in Chapters 4.2, 6.1, and 7.2 to 7.6, we may agree with them.

Larger landholders may have equivocal and varied responses to official attempts at conservation. On the one hand, if erosion is seen to jeopardise the longer-term profitability of the farm, government assistance in the form of agricultural extension, grants for land management works, advantageous pricing policies and subsidies are welcomed. On the other hand, government interference is often resented, and in liberal democracies the 'farm lobby' is frequently quite powerful – in the United States, for example. Well-organised farmers with economic and political resources are frequently instrumental in getting their governments to organise and even partly fund conservation services, as in the predominantly German immigrant farmer community in Rio Grande de Sul of Brazil.

In any concrete analysis, there are four questions to be asked:

1. what precise groups and classes are affected adversely by soil erosion?

2. what power does each of them have in the state apparatus (principally the legislature, the army, police, certain key ministries and their line agencies down to the local level), and outside it - in the countryside, Chambers of Commerce, the shop-floor, and so on?

3. in what ideological terms do these classes or groups perceive the problem of soil erosion - causes, blame, solutions?

4. is the problem of soil erosion perceived to be important enough for them to unite on this issue so that their combined power leads to a coherent response?

Two case studies will illustrate these ideas. It will be argued that the circumstances of Nepal and Zambia are not unusual in that a set of political preconditions for a successful soil conservation policy are not present, and only exist in very rare circumstances (the most outstanding examples being the Republic of South Africa, and South Korea). Usually the classes and groups most adversely affected by soil erosion are politically weak, disunited and spatially separated.

In the case of Nepal (Blaikie 1983), the major impact of environmental degradation is felt by two groups. The first are the poorer urban dwellers – these are largely made up of junior office staff, workers in the retail and hotel sectors, petty retailers and petty commodity producers and the unemployed, old and sick. Quite high proportions of their incomes are spent on fuelwood, and prices have rocketed during the last ten years – from Rs 20/- per load to Rs 50/- in Kathmandu, from Rs 7/- in Pokhara in 1973 to Rs 20/- in 1983. A shift to kerosene is very costly and nowadays fuel for cooking may absorb to up to 15-20 per cent of a poor household's income (see Schroeder 1977, Ives 1981).

The second group are the farmers in the hills, although there are considerable variations between regions and watersheds. Farmers in the *terai* or plains are not really affected except by flooding – although for physiographic reasons most of the flooding occurs to the south of the border with India. It is in the *terai* where large landlords (often absent and in Kathmandu on other lucrative, non-agricultural pursuits), are well established, rather than in the hills. It would be reasonable to guess (although without corroborative evidence) that landslides, gully and sheet erosion, and a general decline in soil fertility would affect the livelihood of poorer and small farmers more critically than the strongly surplus farmer in all areas.

The power that small hill farmers and poorer urban dwellers have in the state apparatus and in society at large is negligible. The most powerful classes in Nepal include the rentier landlord – both urban (in Kathmandu involved in renting houses to shops, aid personnel and in the hotel business), and rural (in the *terai* with large estates, run on a share-cropping basis, and/or with a relative or manager overseeing operations). Then there is the merchant class – discreet and acting behind political figures whom they control with large inducements and dealing in the import of foreign luxuries, and the export of rice, both legally and illegally. Finally there are the contractors, who arrange labour (and sometimes materials) for the multifarious aid projects such as road construction, buildings, and hydroelectric plants. For each of these, environmental deterioration is virtually irrelevant.

The ideological terms in which the affected groups perceive of the environmental crisis are not politically sophisticated – in fact they have very little political content at all. The phenomenon of soil erosion tends to be understood in terms of a mixture of sound and quite sophisticated 'engineering' principles which is implied in a classification of landslides and erosion, and the supernatural. The urban poor have many other issues alongside the escalating price of fuel – particularly imported inflation manifested in high prices for cloth and imported household items, high urban rents and low salaries at the bottom end of the public sector. Furthermore, the causes of fuelwood scarcity must seem remote and diffuse to the average urban dweller. It comes therefore as no surprise that these classes are disparate, weak and ideologically unsophisticated in terms of facing the problem of soil erosion and deforestation.

The impact of foreign aid upon environmental degradation is therefore doubly interesting, and here the need to transcend a crude determinism of economic interests is necessary. The ideological approach to the problem of soil erosion in the international aid and academic communities cannot be 'read off' or predetermined by the relative strength of direct economic interests, as there is often a degree of autonomy of these ideas from economic determinants.

After all, the national interests of donors and the multilateral agencies they finance would be much better served by other projects than soil conservation which is a difficult, diffuse and problematic area for foreign aid with goals that are very long term and difficult to achieve. The interface, therefore, between aid personnel involved in the field of environmental conservation and government institutions in Nepal is interesting and one found in many countries elsewhere. On the one hand, individuals usually on a short-term contract with a job to do face on the other hand a bureaucracy with mixed feelings about the importance of environmental deterioration. The role of rhetoric becomes crucial here – it serves to keep the flow of foreign funds from drying up and to disguise the endemic problems of implementation of these projects.

The case of Zambia offers a contrast to that of Nepal. The population of Zambia is less than half Nepal's, in a land area of more than ten times as large. In spite of low population pressure in an aggregate sense, there has been considerable degradation and erosion in the clay-rich soils which are usually the most fertile, and also in the north where rainfall intensities and runoff tend to be higher (Robinson 1978b: 15). Erosion has also occurred as a result of overgrazing (particularly in the Zambesi Valley), the monoculture of maize and over smaller areas, tobacco, and of some European settler farmers who profitably exploited land in the Eastern Province during the tobacco boom but were not prepared to invest in long-term conservation measures. Yet in spite of a considerable erosion problem, there exists no conservation service as such, with only two ecologists at central level who are supposed to give adequate guidance for land use at the national level. At the provincial level there are Land Use Planning Officers, although their time is largely taken up by the supervision of settlement schemes and in planning state farms (Stocking 1981b). The question arises as to why there is so little official action to combat soil erosion.

The present distribution of political power and the structure of the economy largely derive from Zambia's (Northern Rhodesia) colonial history. Chapter 7.3 examines the colonial impact upon agriculture, and this example confines itself to a very brief exploration of the colonial experience. The area that is now Zambia attracted the attention of British colonisers because of its mineral wealth. Copper, lead and zinc were mined in the centre and north of the country. The economic structure of the countryside near the line-of-rail, linking the mines with ports via South Africa, was transformed to provide labour for the miners on temporary labour contracts in which the miner's family would stay behind on the family farm. Also, Africans were expelled from some of their lands to make way for European settlers who then farmed commercially to sell their produce to the mine workers and other urban dwellers. The

industry, mining &
towns h
power

Zambian economy became extremely lop-sided, in that it relied inordinately upon copper exports. Rural areas and subsistence-level farmers were neglected or actually excluded from the benefits which copper exports made possible in terms of schools, health facilities, agricultural inputs and credit (Bwayla 1980, Klepper 1980, 1981, ODG 1981). Copper mining was resposible for 66 per cent of the country's gross commercial product, 60 per cent of government revenues and 95 per cent of export earnings (Quick 1977: 381).

Since independence was regained in 1964, Zambia has become increasingly urbanised. Indeed, perhaps 50 per cent of the population now live in towns – the highest level of urbanisation in Africa (Simons 1981). The town represents overwhelmingly the focus of political power in Zambia today. First, well-developed mineworkers' unions have been able to influence policy-makers, particularly with regard to pursuing a cheap food policy. Second, the political party which was instrumental in gaining independence and which has governed Zambia since (the United National Independence Party or UNIP) was and still is based mostly in towns, particularly in Lusaka, the national capital. Third, the 'bureaucratic bourgeoisie', inflated in numbers and enjoying large incomes by African standards, paid for by copper exports, is in control of the state apparatus and decision-making (see Szeftel 1980 for a more detailed and sophisticated analysis). In short:

> The elite live mainly in cities and provincial capitals: the seats of government, headquarters of UNIP, the ZCTU and mass media, centres of financial institutions, and the location of major mining, industrial and commercial enterprises. These power centres attract people as honey attracts bees. Operating from them, decision-makers acquire urban habits and perspectives. (Simons 1981: 24)

All urban groups contrive in maintaining cheap food prices–as well as international capital and contractors who have interests in copper mining in Zambia (in order to keep the pressure of union militancy for higher wages from squeezing profits). Therefore rural people have tended to have large surpluses extracted from them through low prices enforced by parastatal marketing boards. Also miners' families are not supported by the 'bachelors' wages' paid to the miner himself. Therefore family farms also cross-subsidise the profits from copper mining. Commercial farmers (increasingly Africans who have taken over in the wake of the slow European exodus since Independence) are still relied upon to provide the bulk of the urban demand for foodstuffs, particularly luxury items such as beef, milk, and cheese. Smaller family farmers have been neglected and offered derisory prices for the products they mostly sell (e.g. cassava and honey, see Chambers and Singer 1980). Soil erosion would only affect these urban interests if it significantly threatened food supplies. Two important factors prevent this from happening. First, foreign earnings from copper have been used to purchase shortfalls in indigenous production. These shortfalls have increased in recent years due to cultivators avoiding cash crops altogether which attract poor prices, bureaucratic and policy-making bungling, and to a general neglect of the building up of physical and institutional infrastructure to aid farming. Second, there is as yet no land shortage, and new land can still be taken into cultivation without much problem. The case of Operation Food Production launched in 1980 is such an example, where enormous state farms have been set up to solve the 'food problem' once and for all. Therefore, soil erosion is a problem faced by small farmers and pastoralists who are politically weak, by large commercial farmers, who can either move on or, in the case of Europeans, leave Zambia altogether. The financial support of the state apparatus and the 'spoils and patronage' system that is a part of it, lies in mining, not in small farmer production. For an introduction to the political economy of Zambia, see Turok (1981).

5. New problems

The first problem after adopting this approach, is the size and scope of the analytical task. From a more conventional approach which encompasses land use, maps of soil erosion, and the designation of technically feasible conservation measures which can be accepted within the constraints operating upon land-users – a difficult enough task in itself – an expansion is suggested to the present programme which seems to involve the political economy of the whole world in every instance of soil erosion. However I would suggest that the problem is more apparent than real.

Certainly it must be admitted that there are a number of relevant issues about development in general on which prior opinions have to be formed – one of the most important is the relationship between rapid population growth and soil erosion which has been briefly discussed in Chapter 2. There are those concerning the nature of the state and the extent to which governments can conserve the environment. Lastly there are those concerning the development of capitalism, particularly whether it is beneficial for the mass of people living in lesser developed countries or whether it could or will become so. Admittedly then, those seeking a comprehensive analysis, will have to have some a priori position on these issues since they all profoundly affect or are affected by environmental degradation. An isolated study of soil erosion and conservation by itself will not provide the answers to these broad controversies. However a bottom-up approach (which always bears in mind the criterion of relevance to land-using decisions leading to environmental deterioration) will leave many areas of local or national economic analysis aside. A journalistic facility for finding and talking to knowledgeable people, reading the right reports and informal commentaries, plus a number of more academic and analytical papers, should enable the researcher following the general methodology outlined here to explain soil erosion in a particular country.

There is a second related problem – that of the terms of reference of the concerned individual for effective action. The question arises who is this analysis for, and who does what about it? This is perhaps the most difficult one and brings up a number of fundamental contradictions. The approach followed here is simply not a policy-orientated one, although it may be helpful to that purpose incidentally. It is primarily an explanatory one, it connects problems and explains why they are problems, and who for, and so does not set up bite-sized problems for the reformer to tackle one by one. It is no coincidence that bureaucrats, foreign aid consultants and ministry officials persist in seeing their role in this manner and trying to solve piecemeal, separate problems. Indeed, without other far-reaching changes in the way people make decisions, this may be the *only* feasible strategy, and this possibility is discussed in Chapter 9. The conclusion reached there is that environmental degradation is and will be frequently beyond the power of the state, and will only become a possibility as an incidental result of other fundamental social changes.

In the light of this general conclusion which will be substantiated from Chapter 7 onwards, it becomes clear that sharpening the tools of policy-making and increasing the 'expertness' of government personnel should no longer be the central concern. This in turn creates one of the most difficult problems of all, a personal one which many government servants, and agents or consultants must feel. It seems almost inconceivable that a project plan, policy document, or even a more abstract methodological contribution on conservation, if well-organised and technically sound, will not have a beneficial impact in the real world. To fulfil the terms of reference for the task, to work hard, and to create a product that is innovative, internally consistent and well-written gives satisfaction. It is very understandable for the person not to want to be aware that it probably will not fulfil its purpose. To blame irrational politicians, political constraints, administrative incompetence, or farmers' traditional beliefs may ease the personal contradiction, but it is analytically naive.

Third, there are epistemological and ideological problems

associated with the bringing together of issues which have hitherto been studied in one or other of the natural or social sciences. On the one hand as was mentioned at the outset of this chapter and in Chapter 2, the physical relationships which determine rates of soil erosion have been examined within natural science disciplines of hydrology, agricultural engineering and agricultural botany. On the other hand the social relationships that determine how people use their environment have been studied in such disciplines as economics, sociology and political science. The point at issue is that both sets of disciplines tend to have different conceptions about the domain and status of proof in the pursuit of knowledge. Whether they *should* share a common epistemology is a debate which it is possible to avoid head-on in this context. The view here is that while the activities of physical sciences, too, are amenable to a social explanation, there are *de facto* different modes of explanation and substantiation. Some of these can be attributed to the subject matter for study itself ('the objects of study are so utterly different that they require fundamentally different methods and forms of explanation and understanding' Benton 1977: 12). This is perhaps believed by practitioners rather than being a view that holds up to epistemologic- al scrutiny. Members of academic, research and other government institutions are divided into two separate scientific communities – the natural and social between which there is little effective cooperation and a lot of envy, disdain and competition for scarce resources.

In the attempt to combine the analysis of physical processes of soil erosion with that of human agricultural and pastoral practices, the expectations of natural and social scientists over the status and domain of proof tend to be irreconcilable (for further discussion readers are referred to more detailed texts such as Winch 1958; Bernal 1969; Benton 1977; Gregory (1978). The main problem seems to derive from the positivist expectation that social explana- tion of soil erosion should be tested in the same way as most natural science ones. While strenuous attempts to subject hypotheses to empirical evidence must be made, the limits of 'proof' must be recognised too. There are always a priori assumptions of what we call an ideological nature, which are not amenable to proof, but which may helpfully be amenable to explicit and prior communication. The assumptions about the state and the world economic system which section 3 above discussed, must lie outside and beyond an effort to prove by means of empirical verification. For example, one can prove beyond a reasonable degree of doubt that a conservation project reduces the rate of soil removal. One cannot prove in the same way that senior bureaucrats do not implement effective conservation because there is no extra financial inducement and it could involve them in embarrassing political contradictions.

6. Family planning and soil conservation programmes – a comparison

As the foregoing critique of current soil conservation programmes shows, there is a broad range of relevant issues about the nature of development itself. Many of the reasons for widespread failure of conservation programmes and policies are the same as for failures in some other development programmes in lesser developed countries. An illustration is given of another important policy issue which has many political and economic similarities as well as the history of its thinking and policy formulation. This is family planning and will be followed through with special reference to one of the world's largest national programmes – that of India. The evolution of the philosophy behind family planning during the last twenty years has moved rather quicker than that of soil conservation so that the present thinking and problems may indicate that in general terms, this critique of conservation policy is not new, but the transfer of ideas merely overdue.

Both family planning and soil conservation are activities which governments attempt to get people to undertake. Both are essentially private activities decided upon by persons within the household or in small groups and involve the deepest and most fundamental concerns in life. Due partly to the propagation and acceptance of various theories of development, (e.g. Coale & Hoover 1958 and Ehrlich *et al* 1973) and perhaps to the direct pragmatic experience of government, states have seen fit to try and persuade, induce or coerce the people to undertake new patterns of reproduction or agricultural practice. One of the major problems with this is that the private benefits accruing to households or families who take up either family planning or conservation measures are often not clear – either as perceived to exist by households themselves or even as calculated by economic models. There is a considerable and divided literature about the private benefits accruing to a household with a small family (Ohlin 1969, saying that children are not a good investment, Caldwell 1978 and White 1976, suggesting a positive asset value for children). Much of the calculation depends upon the choice of discount rate which, it might be added, exists in the mind of the economist (and rather foggily at that) rather than in the minds of a couple about to sleep together. In a similar way many economic models cannot show a positive return to soil conservation measures except on very productive land with high susceptibility to erosion (Harmon, Knutson & Rosenberg 1979: Ervin & Washbourn 1981, for examples in the USA).

In the case of the land-user, implicit discount rates tend to be very high, particularly for the disadvantaged. Fuel for today and food for this season are vital concerns now, and for such people

Keynes' aphorism that 'in the long run we are all dead' may be appropriate. In the case of the couple or family considering the costs and benefits of a future child, the explicit consideration of rates of discount is rather different in that children are perceived to be a positive asset from early on in life but particularly after fifteen years or so when they can work effectively on the farm (and so replace costly paid labour at times of peak labour demand) or as a wage labourer. There is in many societies a strong economic, political and psychic advantage in having male children, so the proposition of preventing *any* births, precludes that possibility, and those benefits are perceived to be real as soon as the (male) child is born. In both cases too, those private choices in some cases bring about the tragedy of the commons (Hardin 1972). Private benefits from more children in the same way as adding an extra livestock unit to the communal pasture are positive, but the public benefits may be negative. Each extra livestock unit added means less pasture available per unit to the detriment of the whole herd. Within a given political economy where there is a 'free' market for labour (taking this, hypothetically as given), labour becomes so plentiful and therefore cheap that it cannot make a living.

Second, there is a different sense in which private individuals may not want to conserve soil or have a small family other than the most obvious that it may not benefit them. Many societies have been conserving the environment perfectly well for millennia, and have integrated conservation into cultivation practices, but not as an explicit programme. Similarly children have been spaced or limited when required by a variety of means (e.g. abstinence after the birth of a child, or delaying the age for cohabitation). Family planning practice as made possible by an operation such as vasectomy or tubeligation, or the use of physical contraceptives makes the prevention of birth explicit and deliberate. Even where religious textual condemnation does not exist as in some societies, there is a strong cross-cultural feeling that such acts are 'against nature' or God. In other words, although people have been conserving soil or limiting children for a long time, an explicit and out-of-context definition of a single element in people's lives (conservation, small families) will tend to be distrusted and misunderstood.

Third, technical problems of conservation and contraception have frequently caused suffering among their unfortunate adopters. Examples of technical failures of mechanical means of soil conservation have already been given in Chapter 5. In the case of family planning, the inter-uterine contraceptive device (IUCD) was launched in India in the mid-1960s before adequate research showed that many women were physically unsuited to it, and it needed a high level of professional judgement on the part of the paramedical staff inserting it. Oral contraceptives too have had long-term effects upon the health of users. Many cases of infections after vasectomy operations have been reported in India (Blaikie 1972, 1975). In both cases inadequate research led to inadequate technical measures being launched and has, in places where the worst effects were felt, put both programmes back decades.

Apart from the reasons why people resist or avoid government-sponsored conservation planning—because of the lack of clear, perceived private benefits—there is a second similarity between the two programmes. The issues 'awakened' by these programmes lie deep in the political economy. A falling birth-rate is brought about by a large number of changes in society. An inadequate list may include reduction in infant and child mortality better employment prospects for educated children making investment in education worthwhile and therefore a cost related to the number of children; women's status in the household and extent of emancipation in making decisions; education of parents and enrolment rates in education for children, and women's participation in the formal labour force (where pregnancy implies unavoidable loss of earnings). To alter these variables, and to a sufficient degree and in sufficient combinations, requires a wholesale change in the fabric of society. In the same way, soil conservation involves an almost equally wide range of social and economic changes. However, in both cases, piecemeal changes with a coherent programme can also have a definite effect. 'Programme' effects can sometimes be isolated statistically, and in the case of family planning have not been insignificant. However, in both cases successful programmes have only occurred when other deep-seated conditions in society have been present which are favourable to reductions in the birth-rate or to effective soil conservation.

For policy-makers, the problems of research and measurement of a large number of important variables, often of a non-quantitative kind, have been severe both in population studies and in soil conservation. Perhaps it might be claimed that the major important variables in measuring population growth rates and their causes are technically easier than those measuring soil erosion rates and their causes. However, as experience in family planning programmes throughout the world has shown, a lack of academic knowledge was not the major problem in achieving their modest attainments, but the inability of governments to shift the major socio-economic variables which encourage the continuation of large families. Some progress was made upon a few, particularly the reduction of major epidemics of malaria, cholera, smallpox and yellow-fever. Others such as the improvement in the status of women, or educational followed by enhanced employment opportunities for children particularly to disadvantaged and poor sections, are much more difficult to achieve (Cassen 1980). So it is with conservation policies.

More research is not a call made usually by conservationists – quite the opposite, in fact (MAB 1979: Holdgate 1982), with the exception of a number of specific technical issues. It has been a lack of progress on other fundamental changes in society that has made effective implementation of either programme so difficult. In the case of conservation, an effective family planning programme (!) and alternative employment for landless and small farmers outside agriculture reduces pressure on agricultural land, pasture and forest. A more remunerative price structure for agricultural products and other measures to raise incomes for the poorest may encourage the use of purchased alternative fuels other than wood or cowdung. It may also help the farmer to bear the immediate costs of conservation measures, and an effective decision-making structure to secure the active cooperation of land-users themselves.

Lastly, both programmes have almost universally been politically unpopular in the countries where they have been launched. Not a single political party in India up to the Emergency of 1975 supported the family planning programme publicly. Likewise many East and Central African leaders avoid conservation policies since these reminded the people of the coercive colonial policies imposed by the British. In both cases, the ideas of a very influential but small number of people in the United States (individuals as well as institutions) backed by large foreign aid funds (and a wealth of expertise, particularly in soil conservation), were responsible for the presence of both family planning and conservation programmes in many lesser developed countries. In an historical analysis of the growth of neo-Malthusianism in US foreign policy, the importance of a high-ranking State Department official, one Philander Claxton Jr, was vital. He was able to ensure that the strong anti-natalist movement penetrated the highest offices of state and was responsible for enormous sums being spent in US foreign aid thereafter (Simon 1981; Warwick 1982). Thus political leaders of lesser developed countries found themselves part of a government that had officially accepted a family planning programme or soil conservation programme (or both), but were aware that there was widespread opposition locally to both.

Thus there are four major areas in which family planning and soil conservation policies are similar – the attempt on the part of the state to change private action where private benefits are not clear; the deep-seated nature of the key or determining variables in the achievement of the policies' objectives; the problems of researching a highly complex field, and the realisation that knowledge of how the key variables operate is not enough, and the fact that these policies are almost universally unpopular with politicians and people in general. It is not surprising therefore that the history of thought about the two policy areas is in some ways similar, although that of family planning is in many ways a decade ahead of conservation.

To return to the case of the family planning programme in India, the 1961 national census provided some very sombre projections of available food supply per caput. The ideology of family planning for the lesser developed countries, plus the formidable apparatus to diffuse it to world leaders was ready, and was translated into American and later multilateral aid for India to set up a nationwide family planning service. The clinical approach to family planning in which medical staff advised upon and provided services at the hospital was replaced by an extension approach in which a vast array of extension agents and network of services in the countryside was set up, using many techniques of marketing and advertising adapted to the Indian context. A policy objective was identified to reduce the crude birth rate in ten years substantially in order to spread the small family norm. Extension services were set up to focus upon a target population of every couple, the wife of which was aged 15-44 years, backed up by medical services and financial incentives to adopters. In this way the private benefits of the small family norm, it was hoped, would be adopted.

There were many problems, predictably, and acceptance was very uneven, concentrated in cities, the more prosperous rural areas, and in literate, middle-class couples (Blaikie 1972; 1975, IBRD 1977a). There were murmurs too that *other* constraints to a small family norm had been grossly neglected, particularly the only slowly falling rate of infant and child mortality in many backward areas. Also the notion that only households with very specific demographic characteristics and specific patterns of access to land find it in their interests to have fewer children gained some publicity (Mamdani 1976).

It was not until the UN Conference on Population in Bucharest in 1973 that a more radical critique of family planning programmes was launched and widely discussed at international level. Also in India, there appeared a number of radical critiques of its own programme which both contributed to and benefited from the conference on population at Bucharest (e.g. Weissman 1970). Here an attack was launched on neo-Malthusianism which was considered by its critics to be politically conservative and which, it was claimed, shifted the burden of blame for underdevelopment to those with the largest families. Rapid population growth was argued to be a result of poverty rather than a cause (and causes were identified elsewhere in the relationship between developed capitalist and underdeveloped countries). The perception of high infant mortality itself was a symptom of poverty and this fuelled a continued desire to have a large family as an insurance against the death of existing children. Landlessness was also seen as an element of poverty and encouraged large families so that children could earn and remit wages. Mamdani

(1976) has put forward analyses which attempt to show that it is in the economic interests of many rural couples not to adopt family planning. In short, rapid population growth was a result of poverty and this asked fundamental questions about access and control over the means of production as well as the technical means by which they were developed. Neo-Malthusianism was further attacked in ways already discussed in Chapter 2, but also drew what in the authors' opinion was a misplaced and utopian analysis from the Marxist literature based solely on a critique of bourgeois ideology, without serious attention being given to the effects of rapid population growth in *any* social formation, transitional to socialism or towards any other social order. (See also Cassen 1976 for a call for more convincing radical population studies.)

The genesis of political consciousness towards conservation programmes arose in very different circumstances, but its development has been moving in the same directions as family planning programmes. Conservation programmes arose both within the United States, and, more or less independently, in the colonies of sub-Saharan Africa. As Chapters 3 and 4 have indicated, a coercive conservation policy for African cultivators existed alongside a voluntary and financially supported one for white commercial farmers. The enforced high population densities endured by Africans due to the expropriation of land for white commercial farmers, threatened to make existing agricultural and pastoral practices very harmful to the long-term productivity of land. The conservation measures put forward by the British and Belgian colonial authorities were highly demanding of labour and moreover symbolised the logic of expropriation and more general exploitation by the colonial powers. While freedom movements in central and eastern Africa were stimulated by these coercive policies, an effective means of ideological diffusion did not lie in the hands of African cultivators nor their leaders, and therefore the analytical lessons to be drawn did not find their way into the international conventional wisdom about conservation. However, there is here an instructive similarity between the two policy areas in question. Both identified a single problem and devised policy solutions without an acknowledgement of the possibility of political change and the removal of the root causes of 'the problem'. Had this acknowledgement been made, both a high birth-rate and environmental degradation would appear symptoms of a wider malaise.

There were two further developments of the family planning programme of India after Bucharest which are relevant to conservation programmes of today. The first was an accelerated attempt to broaden the focus of the family planning programme to include the broader concept of family welfare planning – an ingenious verbal massage of an increasingly tarnished image, but one which did acknowledge the need for a better delivery system for family health including better preventative medicine, pre- and post-puerpueral care and advice on household hygiene which would lower infant mortality rates and improve the health of the family. Other measures were unfortunately not undertaken seriously; unlike better health provision they impinged directly upon the political and economic interests of dominant classes or groups. Adult literacy campaigns (particularly for women), rural employment generation, and land reform or land redistribution which all might have helped to provide a decision-making environment for a small family, as well as other benefits for their own sake, remained at the stage of inaugural addresses, or insignificant pilot projects. A true 'population programme' in the broadest sense failed to emerge.

Conservation thinking has begun to grasp a few of the same nettles. It is now commonplace to read of the importance of local participation, the problem of unequal landholdings, the direct and indirect effects of large-scale commercial farming in the tropics, and 'powerful and vested interests' (see Ch. 5). A few notable articles discussed in the following chapters have taken this further and laid out the implications of successful conservation.

The second development in family planning in India was of a very different kind and involved the attempt by the state during the Emergency of 1975/76 to step up the programme. It appeared as a leading policy under Sanjay Gandhi's Four Point (later Five Point) Programme following the first official policy document (National Population Policy: a statement by the Government of India, 16 April 1976). A highly coercive and administratively effective set of measures were instituted and carried out. In the case of the state of Maharashtra a compulsory sterilisation bill outlining heavy prison sentences for non-sterilisation in certain cases actually reached the State Assembly (*Economic and Political Weekly*, 10 April 1976, p.550). Political will (backed by force – as it had to be in the view of the highly ambivalent public reaction to the programme hitherto) had come with a vengeance (Gwatkin 1979: 29).

During the year 1976 the highest level of enthusiasm for family planning endured, more than 8 million sterilisations were reported, more than three times the number in the preceding year...these meant a nearly 50% increase in the proportion of Indian couples estimated to be protected by modern contraception achieved within a matter of months.

But also, by the time the programme drive came to an end [when the Emergency ended and with it the effective coercive powers of that government], millions had suffered harassment at the hands of government officials bent on implementing it, many, perhaps hundreds, had died from it; the political leaders who had willed it were out of power, and in disrepute, and the programme itself was in total disarray. (See also Banerji 1977; *Economic and Political Weekly*, Editorial, 27 March 1976; and with reference to the backlash effect, Bose 1978)

What is also clear is that it was the weaker and poorer sections of the public that were harassed most. Ironically it is they who have a birth-rate which is falling less quickly, thus 'justifying' that harassment. This is a direct parallel with conservation programmes that concentrate on small peasant farmers and those marginal semi-proletariat which find themselves eking out a living charcoal-burning, cultivating the steepest slopes, or in shifting cultivation without sufficient fallow-periods – an issue which is explored in detail in Chapter 7.

In the case of the conservation policy of India, similar failures to that of the family planning programme have already been noted in Chapter 3. The private benefits to conservation have been extremely ill-distributed, which reflects the underlying dominance of certain classes and groups in Indian government (the industrialists, the capitalist farmers and senior bureaucrats). Between 1951 and 1980 the total area brought under afforestation schemes was 3.18m ha, and 2.2m ha (nearly 70%) were plantations for industrial purposes. The beneficiaries are overwhelmingly large landlords and industrial capital. The beneficiaries are not small farmers except in a few isolated cases where costs and benefits are seen to be equally shared (e.g. in Sukhomaj Village which is discussed in the CSE report, p.14). Second, the cause of the problem has been clearly identified as nomads and graziers who are politically very weak, and who have had much of their livelihood taken away and upon whom the heaviest penalties for 'trespassing' on forest schemes fall – in the same way as in the days of the Emergency, when aggressive family planning 'drives' herded vagrants, petty thieves and other vulnerable groups into camps for more or less forcible vasectomy.

The débâcle for the Indian family planning programme tells us something already clear from conservation programmes under colonial administration. Coercion is possible when there is a government willing to use state power, but it also fails – not solely because coercion as a means to an end tends to fail but because both programmes, in order to achieve any impact, involved very large numbers of people, even the majority, which implies serious problems for the logistics of coercion and of mass political backlash. States that are exploring a tough policy which borders upon coercion like Kenya will either find that open political protest will challenge the power of the state, or that like family planning in India, an onslaught upon the least economically and politically powerful is only temporarily feasible. More passive forms of resistance such as stealing wire and fence posts which exclude people from afforestation projects, ring-barking plantations, setting protected pastures alight, collecting firewood by night and so on, are resorted to and are almost beyond the power of enforcement agencies to stop. Those who advocate tough policies, particularly those who do not seriously question the broad political economic context in which they might be implemented (such as Hardin 1977; Hardin & Baden 1977), underestimate these difficulties and some obvious lessons from history.

In summary, family planning programmes in lesser developed countries since the 1960s have a good deal of light to throw upon soil conservation programmes. Acknowledgement of the failure of targetting upon soil erosion as being *the problem* is long overdue. A full analysis of the causes of soil erosion is required. If a mixed metaphor can be forgiven, Pandora's box remains black, and must now be opened to the light of day.

Chapter 6

Understanding why soil erosion occurs

1. Land use and political economy: a scheme

In this chapter attention is focused on land, land-users and the causes of soil erosion. The approach is a bottom-up one, starting with actual people making decisions on how to use land, and involves a conceptual scheme in which people relate to the environment and to each other. The scheme combines the two elements outlined in Chapter 5 – a place-based concern with where soil erosion is, with a non-place-based concern for political-economic relations between the people who use land, and between them and others. These latter relations are implicit in this scheme and are explored explicitly in subsequent chapters.

There follows a schematic and heuristic device (Fig. 6.1) which suggests the way in which these complex relationships (between people, and between people and the environment) can be handled. As a practical research method it is clear that the amount of data and work required to provide a complete explanation of soil erosion at the micro-scale (say, 1:25,000) is enormous and usually impracticable to carry out in its entirety. However, judicious use of aerial photographs and sample surveys of land-users provides a combination of locational and social data which may make most elements of this model a feasible research undertaking. Some illustrations of parts of this scheme are offered, but since it is new, extensive in scope, and combines a very broad spectrum of geographical scales and levels of specificity, it is hardly surprising that the whole scheme cannot be illustrated by a single case study.

The scheme can be likened to a watch, a reliable and easily understood mechanism to keep time. What follows is the explanation of how the watch works, and its design is at least on test in the sense that it can be wound up and its accuracy tested (a specific study undertaken and its ability to handle a complex reality put to the test). However, the theory of time, the use to which the watch is

Note: This chapter is a modified version of material used in Blaikie, P.M. (1981) 'Class, land use and soil erosion', *ODI Review*, 57–77.

Fig. 6.1   Land use decisions and soil erosion

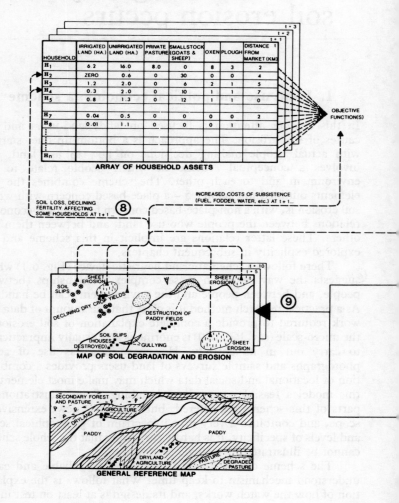

ARRAY OF HOUSEHOLD ASSETS

MAP OF SOIL DEGRADATION AND EROSION

GENERAL REFERENCE MAP

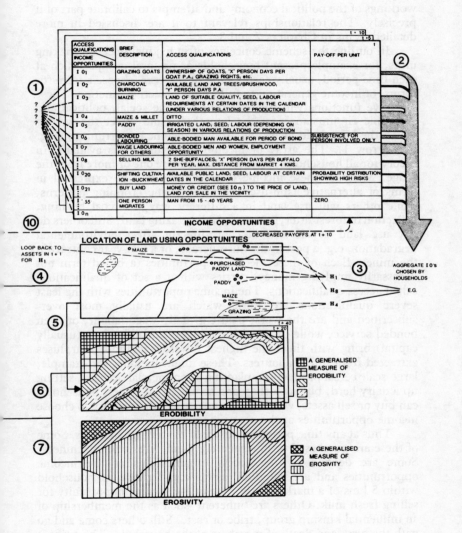

put, the meetings that must start punctually and so on, all lie outside this device. The watch needs a time but not symmetrically vice versa, such that no one would expect a watch to give its user any clue on how to use time. In the same way, this scheme derives its whole purpose from the political economy itself – it merely maps the workings of the political economy and attempts to calibrate part of it precisely. The relationships relevant to it are discussed in more detailed terms in Chapters 7 and 8.

In outline the scheme conceives of individual decision-making units, for example peasant households or pastoral groups which at each time period, usually each season of the year, choose a form (or forms) of income generation to fulfil some objective function. These objective functions will vary according to the social, political and economic circumstances of the household, and affect the way in which the household views the array of income opportunities (Fig. 6.1, arrow 1). The array of income opportunities is available in theory for all households, each of which choose one or more from the range. Many of these income opportunities will be expressed in terms of alternative land-uses such as specific cropping patterns, while others will use land in a different way, charcoal burning, collection of forest litter, or use of public grazing land. Still others do not use land at all and do not directly affect environmental degradation, e.g. a post in the bureaucracy, or a two-year contract in the mines. Each of these income opportunities (a broad term not necessarily implying money) carries with it a set of qualifications called access qualifications. The income opportunities with the least severe qualifications are those which are usually most over-subscribed and are the worst paid e.g. daily wage labour, or even bonded service, while those with the highest qualifications (usually capital) bring with them the highest rewards, including surpluses extracted from hired labourers. These might include, for example, large-scale cereal production, planting a commercial orchard, setting up a dairy herd, buying a shop and so on. At all stages, households can buy or sell assets, which will in turn affect their ability to choose income opportunities at the next time period.

Thus at any time period a household has a range of assets. Some of these are purchased or inherited e.g. land, livestock, machinery. Some are defined by the access qualifications of the income opportunities and are 'windfall' assets e.g. location of household within 5 kms of a market in the case of the income opportunity for selling fresh milk. Others are 'inherent' such as the membership of an influential kinship group, tribe or caste. Still others come and go with the birth and death of members of the household. The point is that they can change through time as the fortunes of a household wax or wane.

The analyst has a choice over which decision-making units to include in the model. Usually all units which use land in an area will be included. Thus the majority of these units may be households, but there may also be a logging company, a state farm, or a private plantation, and naturally the array of assets will be rather different as well as their objective function. Thus, when the word 'household' appears in this explanation, it is acknowledged that there may be other decision-making units which are not households at all.

Diagram 1 in Fig. 6.1 illustrates schematically the list of assets for each household at each time period, and the individual strategies which each household follows, the outcome of which is fed into the assets array for t + 1 (usually in terms of accumulation and investment, subsistence with a balanced household budget, or disinvestment). The latter course has to be followed to meet minimum consumption requirements, and if there is no other asset to sell, a loan must be raised or migration of one or more members of the household contemplated.

The way in which the objective function(s) of households (or other groups) in this scheme is defined by the analyst should be judged in the light of its pay-off in understanding land-use decisions, and in the ability to distinguish between those income opportunities which are potentially soil eroding and those which are not. It does not necessarily require a very sophisticated and highly quantified objective function to be specified. If the data are available, some sensitivity analyses could be carried out using a small data set (say a dozen farmers or a sample from different types of farmers). If they are not available, a more qualitative assessment of what objectives land-users have may be adequate, perhaps with some issue-specific calculations on the effects of policy changes, population growth, or yield expectations, which might yield sufficient insights into the causes of erosion. In a version using computerised simulation worked through by Blaikie, Cameron, Fleming and Seddon (1977) the objective function was defined as follows:

> The household scanned all the possibility routines (income opportunities) open to it according to its access profile, and adopted that in which earned it maximum income (all income opportunities having been converted to a cash value), and to the maximum extent it could, subject to constraints. With the remaining resources left to it (not committed to the first opportunity), it took the second most lucrative, and so on, until it could not adopt any more routines.

The maximisation of income, and the expression of income as a cash value, may not seem an accurate approximation of peasant or pastoralist behaviour. However, a more complex objective function could easily be substituted here, and this one described is not central to the logic of the scheme.

To take an illustration of a very different objective function, in many parts of southern and central Africa cattle are looked upon as a store of wealth rather than a source of income (Doran, Low & Kemp

1979, Stocking 1982). These authors have noted that there are large areas of overgrazing in Uganda, Tanzania and Kenya and destocking is urgently needed in certain districts. The objective function of cattle owners is to accumulate and conserve cattle assets which confer security, prestige and status. The cash value of animals is important only so far as current consumption is concerned. Doran *et al.* follow through in the Swazi case the logical implications of this view of cattle and test their hypothesis by regression analysis. Most of the efforts to set up cattle fattening ranches, breeding stations and pasture improvement were predicated on the assumption that pastoralists would wish to reduce stocking densities, since the same or improved incomes could be generated by fewer cattle per owner. This effort failed and herd size *increased* over the period 1968–77 because of the objective function of cattle owners. In this example it is important to define the objective function very carefully. Again, it is up to the analyst to be in a position to know just how sensitive the approach is to changes in the specification of the objective function.

The class composition of the society under study is therefore, in a sense, 'operationalised' in terms of the income opportunities and their access qualifications. Each one has implied relations of production, forms of surplus extraction, investment and accumulation possibilities. Thus households are impelled through history, and the general processes of population growth, differentiation, and class formation can be charted in successive iterations of the scheme through time (see Caxner 1982, for a similar scheme in the Philippines).

It could be argued at this juncture that this part of the scheme suggests that households only act individually, and that it precludes all recognition of struggle (not individual but collective action which, if politically successful, could change the dominant relations of production and hence the whole structure of income opportunities and access qualifications). However, these processes can be easily accommodated by altering the 'rules of the game', which are themselves defined by the political economy at large. An example of a change in the 'rules of the game' would be a radical land reform programme as was undertaken in China, and with shorter-lived effects in Peru in 1964. Another less fundamental one might be the widespread formation of cooperatives (e.g. in Zambia from 1965–69 until the movement collapsed). Hence changes to households could occur by a complete redefinition of household assets and through investment and disinvestment and depreciation from season to season, or in changes in the availability of income opportunities (some appear, some disappear from time to time); and in their access qualifications (due to significant shifts of power in a society, but also through forces of the market, and, as we shall see presently, through secular changes in the environment).

It now remains for the scheme to be placed firmly within a locational framework and to unfold within a particular landscape. Without a locational framework for land-use decisions, their implications for environmental degradation cannot be understood. In terms of the two paradigms discussed in Chapter 5, it is at this point that political economic relations are superimposed upon, and find geographical expression at, a specific location. Sometimes the extent of soil erosion is related by quite subtle and seemingly minor characteristics of agricultural practice. For example the time of planting of crops can be important in determining the degree of protection the plant might give to bare ground at the time of maximum severity of rainstorms. Whether ploughing is undertaken accurately along a contour or down the slope, the spacing of plants and so on are other examples of these characteristics. Clearly these are not 'explained' by the scheme, although in *some* cases specific access qualifications and assets combine to explain a specific aspect of agricultural technology. For example, the scheme could explain the existence and distribution of mechanisation in agriculture, but would only provide a context within which variations in other agricultural practices occurred. For the purposes of identifying soil degrading activities, certain income opportunities (usually crop combinations) and the households which undertake them can be asterisked and specially noted to be damaging to the environment in areas of high erosion hazard. The farming systems approach is not incompatible with this one, in that it identifies groups of farmers with similar agro-ecological context and control of assets, and also relates non-farm activities to land-use decisions.

Figure 6.1, arrow 2, shows how various aggregate land-using decisions are mapped. The thickness of the arrows leading out of various income opportunities indicates the relative frequency each income opportunity has been taken up in any one season or year. Now the location on a map of these decisions may well show marked spatial clustering for two reasons—the first concerning the clustering of people of a similar type, an issue of 'eco-class', and the second the clustering of income opportunities, and these reasons are explained below.

First, groups of households with similar access profiles are sometimes spatially marginalised, and are obliged to carry out their land-using income opportunities in similar 'marginal' places (areas of poor soils, steeper slopes, riskier and scantier rainfall etc.). In Figure 6.1 three vulnerable households (H2, H4 and H8) own very little private land and, therefore, to reach their minimum consumption requirement have to labour for others (income opportunity IO7) and eke out a speculative additional income from cultivating buckwheat on very steep slopes (IO20). On a larger geographical scale of resolution, whole tribes or peoples (such as on the Trust Lands and

Native Reserves of eastern and central Africa, or even whole nations such as Lesotho) are spatially marginalised, and in certain instances populations are concentrated by exclusion, or by ownership of land by others. An extreme case is the Bantustans in which population pressure leading to environmental decline has been a constant problem (see Beinhart 1981). Secondly, specific income opportunities (which involve land use) are either to be found only at a specific location (e.g. paddy or irrigated land) or are more profitable at one location because of distance criteria, suitability of the soil for a particular crop, and other considerations which might be labelled as locational comparative advantage.

Thus the land-using decisions of the rural population can be accumulated over a period of, say, five years and mapped at an appropriate scale – 'appropriate' referring to the level of generality of the overall analysis. If a detailed study of environmental degradation is being made, showing how particular households' decisions were adversely affecting particular areas of a single watershed, the scale might be as large as 1:25,000; while the approach could be broadly similar in the investigation of the effects of the impoverishment of 100,000 Kenyan farms on Class III and IV lands to the north of the White Highlands, in which the scale and level of detail would clearly be substantially reduced.

Another set of maps showing the spatial variations of the physical determinants of erosion are necessary, and are in essence 'laid over' the map of cumulative land-use decisions. Figure 6.1 shows maps of erosivity and erodibility, but there are a large variety of other ways of measuring and mapping the vulnerability of soils to degradation and erosion (Stocking and Ellwell 1976; Morgan 1979). The way in which this essential element is measured will also depend upon the geographical scale of the analysis. In Figure 6.1 the two maps of erosivity and erodibility together, and the other of accumulated land-use decisions, provide the basis for an immediate and directly causal explanation of environmental degradation (arrows 4, 6 and 7). Erodibility is affected by land use as well as physiographic factors and therefore the cumulative effect of land-using opportunities will change the erodibility status of the land (see arrow 5). Since it is difficult to attribute accurately the effects of people upon soil erosion, this approach therefore suffers from the same weakness as any analysis which attempts to assess or attribute soil erosion in some degree to specific land-use decisions, rather than to causes that could have otherwise operated under different conditions. Thus it is quite likely that arrows 6 and 7 will have more explanatory power of the spatial variation of soil erosion than the arrow 4 (and the whole analytical framework behind it, which leads back to the access profiles of the rural population).

The element of geographical scale is one taken up by Gardner

(1981) both as a more technical issue with regard to FAO's Soil Degradation Map, but also showing how physical and political economic manifestations exist 'on the ground', and has been discussed already in Chapter 2. As has been mentioned, the scheme presented here is not scale-specific, and at the most detailed and small-scales (of, say, a few km²), the political economy may even have a geographical expression here too, although it is my experience that spatial variations or erodibility usually have a much greater explanatory power on the exact location of soil erosion or degradation.

There is one final point regarding this scheme which concerns the feedback effects of both (a) decisions taken by households upon their future decisions (about the next season or the next year and thereafter) and (b) environmental deterioration upon both the households' assets themselves and the access qualifications of certain income opportunities (in Figure 6.1 the dotted arrows 8, 9, and 10 linking the map of soil erosion at t, with the asset array and income opportunities at t + 1 and t + 5 respectively). The latter feedback effects are an example of the specific working out of dialectical relationships between people and nature. To take some illustrative examples, environmental decline causes a fall in yields (arrow 10) for both agricultural and pastoral income opportunities, and therefore reduces the income of certain (or all) households (arrow 8). Some of the more advantaged will be able to diversify by meeting the access qualifications of other income opportunities or by merely retrenching and reducing consumption levels (arrow 9). Other households may not be so lucky, and the decline of their fortunes may precipitate them to the status of semi- or complete proletarianisation. This in turn may force those households to use the public economy more intensively as the only escape from working for others (and thereby exacerbate the 'tragedy of the commons' in specific circumstances as outlined later).

Looking at these feedback effects from a different perspective, the access qualifications of different income opportunities also become more severe. The additional labour required to collect fodder from the forest to maintain a buffalo, or to collect wood for charcoal burning, may be due to declining forest resources and will increasingly exclude those households which are short of labour. The daily tasks of fetching water and fuel will become more onerous, raising the fixed cost of subsistence and reducing resources for allocation elsewhere. There are well-documented cases of the rising cost of fuels, particularly around towns (e.g. Digerness 1979 in Sudan, Eckholm 1976 in Niger, Chauvin 1981 in Upper Volta), but also in rural areas where fuel demands are high. In Niger, Eckholm (1976) says that up to a quarter of rural households' income is spent upon fuel. Whereas a generation ago, Nepalese hill households spent one or two hours to collect woodfuel, it now can take the labour resources of one person continuously throughout the year to collect sufficient or at least a biannual expedition of a fortnight or more. Usually this affects the poor much more than the well-off (Briscoe 1979, Douglas 1982 for Bangladesh; Agarawal 1980: 14; Mnzava 1981 for Tanzania), part of the reason being reduced labour availability to work for others, or the inability to employ others on their own farms because of the necessary allocation of labour to collect fuel (Ashworth & Newendortter 1982). Similar examples exist for the increasing time taken to fetch water where perennial sources have dried up. In these ways, peasant differentiation may be accelerated. A general process of impoverishment, alongside others to be described in the following sections, all encourage a more desperate, less finely-tuned use of the environment, where there are simply few resources to be allocated to long-term consideration of soil, pasture and food conservation, and such natural resources as are still available have to be used immediately in order to survive. Also this feedback effect has proportionately more effect upon women *within* households (e.g. Dhogra 1980; Ki-Zerbo 1981; Nagrobrahman & Sambrani 1983), an aspect not modelled in this scheme.

In the next chapter we turn to the more general processes of impoverishment of many land-users in lesser developed countries and by so doing make explicit, and provide an analytical framework for, the political economy of the contents of the scheme, whatever they may be in a specific context. The framework will seek to provide an explanation of: (a) distribution of assets across households (or other decision-making groups) and the ways and directions in which they change through time (e.g. processes of impoverishment and differentiation); (b) the range of income opportunities which reflects the nature of the economy. This may be underdeveloped and dependent, for example, which would be reflected in a preponderance of lowly-paid jobs with little choice, and often located far from workers' homes; (c) the access qualifications of different income opportunities, which variously describe the costs of inputs, including capital, land and labour, possible discrimination on the basis of race, caste, tribe or kinship, and so on; (d) the pay-offs of the income opportunities which are determined by such considerations as the level of wages; yields from different crops; the rate of capital accumulation and the various means of surplus extraction from productive labour; the prices of products (which may be different depending upon the contractual strength of the seller versus the merchant), and so on. These are only some examples of the way in which implicit social, economic and political processes, which show up in this scheme as merely an array of numbers or categories, can be identified. The later chapters attempt to provide a method of analysis for these processes.

Chapter 7

The exploitation
of natural resources
and labour

[handwritten: ∠ peasant majority]

1. Introduction

The scheme in the previous chapter related individual choices of land-use, the distribution of those choices among various land-users, and their actual location on the ground. Necessarily the scheme was mechanistic, but it was also conceptually precise. In this chapter we seek to 'animate' the scheme with an analysis of the political economy. The analysis of soil erosion taken as a whole must embrace *both* the mechanisms of choice of land-use as well as the more abstract theory which explains those choices, who makes them and why. It is to this latter concern we now turn.

In this chapter we look at small farmers and pastoralists who make up the majority of land-users in lesser developed countries. They are connected in a great variety of ways to other elements (groups, classes and institutions), through the market, laws, struggle (both political and/or armed), and even rural development projects. Therefore in this chapter, the other groups have to be specified too. In Chapter 8, attention moves to these other (and usually more powerful) elements such as transnational companies, plantations, logging companies, state farms, and such like.

2. Land uses, social relations of production and exchange

A central assertion in this book is that soil degradation and erosion directly result from cumulative land-use decisions through time and that these decisions must be considered as a part of a wider political economic analysis. This section starts with the relations of the decision-makers themselves and other groups or classes at the level of the enterprise itself (i.e. 'on the ground', at the point of production). In the following section those immediate relations are put in the context of the world economic system. The major focus is upon the peasantries of lesser developed countries and their relations between themselves and with the world economic system in so far as these relations explain soil degradation and erosion. Here two major spheres of political economic relations are identified which are crucial in understanding land-use decisions and soil erosion: the social relations of production at the level of the enterprise (the farm, plantation, ranch, herd, etc.), and exchange (the prices or costs of production and the product).

In both these spheres, surpluses are extracted from peasantries and pastoralists, at the level of the place of production or enterprise. For instance, peasants may be obliged to work as wage labourers to make up shortfalls of staples and cash requirements for household goods, as well as to pay for implements and other inputs for the production process itself. Employers in this instance tend to accumulate and the 'peasant as wage-labourer' works to create a product which is sold by the employers at a value which is more than what it costs to replace the physical capital resources used and the wages paid for labour (Deere and de Janvry 1979). Rents in kind or cash are other typical means of appropriation.

The second sphere in which an explanation of soil erosion may be found is that of exchange. Here again it is important to relate the explanation to particular relations of production. Surpluses are extracted from peasantries, not only at the point of production (when peasants often work for part of the time as wage labourers), but also by terms of trade unfavourable to them. Low agricultural prices may develop for many reasons such as the presence of monopolistic merchants buying cheap; food policies which give low prices to producers, and competition with large capitalist producers, or 'transitional-upward' peasants. Indebtedness, due to the instability of peasant production and the prices small producers get for their produce, and their usually small capital reserves, is often also a mechanism for surplus transfer. Lastly taxes, imposed by the state often are in effect a transfer of surplus from the peasantry to other classes.

Bernstein (1977, 1979), in two important papers on capital and the peasantry, highlights what is called the simple 'reproduction squeeze', by which is meant a result of deterioration in the terms of trade between necessary purchased commodities for consumption and agricultural inputs and commodities (usually primary agricultural produce) sold by peasants. This leads either to a reduction in consumption or an intensification of commodity production. The costs of production by the peasant are raised, leading to exhaustion of land and/or labour. The exhaustion of land and labour go hand-in-hand as more labour time is required to work poorer, degraded and more distant soils and so returns to labour are reduced.

3. Land users and the world economic system

The social landscape (as well as the physical, in many instances) in which the peasantries and pastoralists find themselves in lesser developed countries at the present time has long been affected by European economic, social and political influences, first in the form of European traders, and later of imperialism, as part of a longer-term development of world capitalism. It is not the concern of this book to chart the historical development of capitalism and to enter into wide-ranging debates on imperialism, but to focus on the ways in which the development of capitalism affects peasantries and pastoralists, and thereby the ways in which they use the environment.

In general terms, it is widely agreed that, during its historical development, capitalism has had different requirements from the peasantries of the world at different times – raw materials, land, labour power, and at times of crisis, markets – and that the means by which the self-sustaining peasant economies are rendered malleable to these needs is through the introduction of cheap commodities, through military and police force, and through the state (including the imposition of taxation and restrictions, laws and other policies concerning agriculture, mining, etc. Luxemburg 1963, Bradby 1975). In the case of South Asia, British imperialism was additionally concerned to extract revenue from the land in the form of land taxes.

The requirements of different imperial powers varied, and also changed through time, and a short account such as this cannot do credit to the complexities of local histories. A few examples of the relationship between the ways in which raw materials, labour power, land and so on were 'shaken free' from the societies which originally used or occupied them follow; the implications for land use and soil erosion are taken up later on.

In Zambia, for example, the promise of minerals, particularly zinc and silver and later copper, brought British interests into the area from the south in the form of the British South Africa Company. The problems then arose of recruiting labourers to work in the mines, and secondly of feeding them. The labour demand from the mines in southern Africa was met by a combination of taxation (house and poll taxes which had to be paid in cash, thereby requiring a cash income) and eviction of African cultivators from certain areas which were then designated Crown Lands and reserved for settlers (Cliffe 1978; Klepper 1980). The problem of cheap provisioning of the mine workers was solved first by encouraging African producers to grow food for sale and then by discriminating against them when eventually white settlers (in fewer numbers and much later than hoped for) arrived and established large commercial farms near the mines and along the line-of-rail. Hence displacement of African cultivators to make way for European settlers, labour migration from the African reserves to the mines, and the extraction of minerals were combined in a particular fashion and can be seen to have changed through time. A general process by which capitalist relations of production became dominant is quite clear - both in the mines themselves and on settler farms. However, the peasantry was left *partly* in control of its means of production so that it could supply the mines with cheap food when it suited the colonial interests of the time. The mechanisms of surplus extraction outlined in the previous section can be readily recognised here.

The implications for land-use were as follows. First, much higher population densities of African cultivators rapidly occurred, causing soil degradation and erosion in some areas. As stated by the Department of Agriculture of Northern Rhodesia in 1936 about the Reserve Lands around Chipata: 'Where the population is dense, many of the once fertile valleys have been exhausted by over-cropping and erosion and the villagers have taken to the cultivation of incredibly steep hill gardens.' Increases in stocking rates in Reserve Lands also accelerated the erosion rate, perhaps even up to a hundredfold, although detailed measurements are not available (Robinson 1978b). Frequent mention of other reserves facing similar problems can be found; for example, the Chewa and Ngoni Reserve Lands in Eastern Province, the Petauke and Fort Jameson Reserves in Southern Province, and so on (ibid: 32).

Second, shortages of labour occurred at critical times during the agricultural calendar, leading to both differentiation among the peasantry and increased vulnerability of certain households whose prime labour resources were absent from home (e.g. with women as head of household). In turn this modified certain practices, particularly the *chitamene* system, causing a decline of soil fertility (Allan 1967: 128). Thirdly, inmigration from neglected outlying areas to the line-of-rail, both to mining towns and for settlement, increased population pressure in these areas. Finally, there existed a preoccupation on the part of the Department of Agriculture with research into and encouragement of cash crops to feed the town and mine workers. The crops were predominantly maize, and (although to a lesser extent than in Kenya,) tobacco and cotton for export, to the neglect and implicit discouragement of the development of peasant staples such as millet and cassava.

The combination of these rapid and enforced changes in the socio-economic fabric of the area had a profound effect. Of course, the movement of peoples throughout Africa had long predated Arab or European interference: adaptations of agricultural and pastoral practice had usually occurred without severe environmental damage, and pre-colonial soil conservation measures have been fairly widely acknowledged by Western writers (Ranger 1971). However, unprecedented colonial demands for labour, tax revenue, economic crops and land led in Northern Rhodesia, where there were relatively low population densities even for tropical Africa, to quite widespread environmental degradation (Kay 1972). Not all areas succumbed to these pressures, and an impression of uniform and universal subjugation would be misleading. Various characteristics of the pre-colonial society itself either accelerated or slowed down the impact of the colonial power, and thus the economic penetration and domination of those societies as Hyden (1980) has pointed out about the peasantry in Tanzania, and Rennie (1978) for the Ila economy in Zambia, together with the well-known and meticulous anthropological study by Watson (1964) on the Mambwe. Furthermore, various peasant struggles against the colonial powers were widespread, although not outstandingly so in Northern Rhodesia. One aspect of these struggles was a strong element of distrust and resistance to soil conservation measures imposed by the colonial administration throughout British East Africa.

While there were a number of significant changes in the political economy of Zambia following independence in 1964, the fundamental aim of the Government's agricultural policy continued to be cheap and reliable food supplies for the towns. As state marketing systems developed, peasants in outlying areas were compelled to grow maize and sell it at prices that usually did not cover cultivation costs for small farms except in favourable areas and in good years. Although there was a slow exodus of European farmers, land was not expropriated since they were seen to be essential to the provision of food (and indirectly to the holding down of mineworkers' real wages, thus increasing profits for the Zambian Government). However, new and more capitalised African farmers (termed 'emergent' farmers) were also encouraged by means of credit, subsidies on inputs, and an extension service. Hence many of the characteristics of colonial agriculture were preserved. In this the 'bureaucratic bourgeoisie' were in constant contradiction with the party (UNIP), the former always advocating policies which favoured larger farmers, and cheap marketed surplus (and blocking alternatives which would take resources away from achieving this end), and the latter a more socialist policy of reaching and benefiting the small farmer. The former has, by default and active politicking, prevailed throughout the many shifts and turns in agricultural policy since Independence (Quick 1977). As a result the characteristics of post-independent Zambia are:

(i)   a sharp regional differentiation between line-of-rail and other areas, and between the more commercialised Central, Southern and Eastern Provinces and the rest;

(ii) a continuing heavy emphasis on commercial crops, and research into crops, inputs and practices for the larger commercial farmer;

(iii) continued and increasing rural-urban migration, partly due to neglect and continuing extraction of surpluses from rural areas and ensuing poverty. (See Marter and Honeybone 1976, and Chambers and Singer 1980, for good accounts about differentiation of producers and some of the problems of poverty which many of them face.)

Thus most of the original processes which had led to soil degradation and erosion continue. Robinson (1978b) quotes the words of the Land Use Services Division of the Ministry of Rural Development in its Annual Report of 1974 'Soil erosion is proceeding unchecked in many localities and gives cause for concern. This year's heavy rains served to emphasise the problem both on State Land Farms, and in the Reserve and Trust Lands.' The effects of the various processes of peasant differentiation and government support of commercial farmers upon precise land-use patterns and agricultural practice, and thence upon soil erosion have been analysed by Stocking (1981b, 1983). The contrast between the farming systems of a commercial type on the one hand and more subsistence-orientated mixed cropping systems on the other is striking in terms of annual rainfall interception and erosion index. Each system is associated with a particular size of farm – small scale (subsistence), 'emergent' or medium (commercial) and large scale (commercial) – the latter two groups attracting almost all government assistance and the former group suffering direct and indirect discrimination at the hands of government policy (both intended and by default). The earlier study (Stocking 1983: ibid.) is one of the few studies that follows through political and economic structures at the national level right down to soil eroding and soil conserving practices on the ground.

This very abbreviated account of the colonial impact upon land use and soil degradation and erosion in Zambia illustrates a particular colonial experience. Some of its aspects are common to other southern African states (particularly with regard to labour migration), while others further north, apart from Uganda and parts of Tanganyika, shared the broad aspects of displacement to make way for white settlers and various sorts of discrimination against African cultivators. (For a more general account of sub-Saharan Africa's imperial experience, see Rodney 1972; Brett 1973; Palmer and Parsons 1977). The situation described in Zambia is a very common situation in which peasants work for others in mines, estates and commercial farms for part of the time, since the produce from their own farm yields insufficient to support the household.

Two examples of rural development projects in Latin America given by de Janvry (1981) illustrate well the relationship between peasants and capitalist farming, and also the implicit political objectives of rural development projects, when the results of that relationship threaten the status quo. The first example is the Puebla Project in Mexico where:

> ...the land reform of Cardenas had created an efficient commercial sector,

concentrating the bulk of the land in commercial agriculture and assuring rapid growth in the production of wheat and exportables; at the same time it had settled the mass of the peasantry in the *ejidos*, thereby creating a source of cheap food and the labour reserve for commercial agriculture. As erosion and demographic explosion increasingly threatened both the delivery of corn surplus to the market and the social status quo after land reform, public policies directed at reproducing the conditions of functional dualism were sought. (ibid.: 234)

This quotation illustrates three issues. First, the relationship between peasant and capitalist enterprise is one in which the peasant frequently *has* to work for part of his time in capitalist enterprises. The means by which this comes about is by capitalist farmers pushing the peasant off his land *so that* he is sub-marginal and is looking for a subsidiary income – which has similarities to the Zambian case just outlined. Second, there is the issue of functional dualism; the peasantry's farming operations thereby subsidise commercial operations since the latter would have to pay full-time wage labour the full cost of its physical reproduction (i.e. enough to support the labourer *and* his or her family). The family left behind on the peasant smallholding can provide for itself and therefore a 'living wage' for the single worker and his family need not be paid by the mine or plantation owner. When demographic pressure and erosion affect these holdings, therefore, it is *not only* the pocket of the peasant which is affected. An example in the Republic of South Africa and its African Reserves between 1930 and 1960 is a case in point where 'the menace of soil erosion' threatened, with induced population pressure, to undermine the viability of the labour reserves causing breakdown and starvation. There followed an interesting battle between the Department of Agriculture which wanted to create independent peasants in the reserves, and the Department of Mines which wanted to keep them dependent and therefore available as a reserve labour force. Lack of land and other political pressures ensured the Department of Mines won the day (Beinhart 1981). The third issue is one of spatial marginalisation in which peasants have to use marginal land, and this is discussed in more detail in section 5. The Cajamarca Project in Peru (de Janvry 1981: 237) illustrates this point in which 'these haciendas coexisted with a large number of subsistence farms devoted to the production of grains and other staple foods on the steep, dry and eroded slopes of the valley's flanks'. Indeed, the haciendas expropriated numerous of their *colonos* and consolidated their holdings into large pasture operations when new commercial opportunities (a road plus the establishment of a Nestlé plant) appeared.

Another detailed example of the relation between the peasantry and the large farm in the Cauca Valley, Peru is given by Taussig (1978). Here both permanent workers as well as contract workers are employed in large estates, but also have their own farms, sometimes at a great distance, and move between two-year contracts in the estate and periods in between back at their farms. Here too are the historical processes of dispossession of peasants, rural development projects aimed at forestalling occupation of the estates by peasants, and government-sponsored monoculture resulting in an increase in erosion and flooding. The problems of peasant agriculture, including environmental degradation, are seen as resulting from the existence of the estates, and a government which broadly represents the class which owns them. Other examples in Central America in which transnational companies have been the other partner in functional dualism are commonplace (e.g. *The New Internationalist*, 1982, Feder 1977).

These examples (and there are many more, broadly similar) must not give the impression that *all* peasants and pastoralists are related to capitalist enterprises in this way. There are many peasantries that have a much more tenuous and indirect relationship. The example of the Nepalese peasantry which appears in this book a number of times is a case in point. Although there are outmigrants who serve in the Indian and British armies, work in tea plantations and as guards, porters and menials throughout India, the peasantry as a whole could hardly be said to be in a functionally dualistic relationship with capitalist enterprises. Remittances from outmigrants are important and do 'prop up' the failing hill economy, but there is little evidence of a widespread and direct relationship with capitalist enterprises, and certainly the latter could do without, and is economically indifferent to Nepalese workers. So, in many other parts of the world the functional dualism of Latin America and the labour reserve areas of central and Southern Africa does not operate in the same close and direct manner as described by de Janvry (1981).

4. Marginalisation, proletarianisation and incorporation

As we have said, one of the fundamental theoretical assumptions of this chapter is that soil degradation and erosion can be explained in terms of surplus extraction through the social relations of production and in the sphere of exchange. The essential connection is that, under certain circumstances, surpluses are extracted from cultivators who then in turn are forced to extract 'surpluses' (in this case energy) from the environment (stored-up fertility of the soil, forest resources, long-evolved and productive pastures, and so on), and this in time and under certain physical circumstances leads to degradation and/or erosion. We shall now examine these processes in a little more detail. There are three interlocking (and in some usages overlapping) concepts which describe them: marginalisation, proletarianisation and incorporation. The literature is very large indeed on all of them. A lack of space, the issue of relevance to soil erosion, and my partial reading of it, will have to suffice as excuses for a very brief treatment of these terms.

Used in the context of peasants in lesser developed countries, marginalisation has tended to imply the process by which they lose the ability to control their own lives (where they live and derive their income, what crops or stock they produce, how hard and when they work). This comes about because of their incorporation into the world economic system. Presumably also by extension the meaning of marginalisation can also imply that they can be incorporated into pre-capitalist modes of production or into enforced collectivisation in the Soviet Union, although I have not seen the term used in this context. Implicit in this process of incorporation (poll and hut taxes, coerced migration to mines, and soil conservation measures have already been illustrated in the Zambia example) is its significance for changing relations of production and the process of proletarianisation. This means in the peasant context the loss of independently controlled land, livestock, implements and direction of its own labour by the peasant household itself, and the enforcement of some sort of wage labour for others who own or control the means of production.

There is an additional development to the concept of marginalisation which is particularly relevant in the study of soil erosion. It was Wisner, O'Keefe and Westgate variously and together who extended this idea in a very simple and effective way: for example, '...in such a way socio-political and economic marginality produces eco-demographic marginality, i.e. *marginal people* are, through the process of social allocation in the neo-colony, quite literally pushed into marginal places' (Wisner 1976).

A well-integrated case study (Franke & Chasin 1981) entitled 'Peasants, peanuts, profits and pastoralists' relates the penetration of agribusiness into the economy of Niger, reducing food security and incomes for pastoralists and many farmers alike, and causing erosion and pasture degradation both indirectly by the spatial displacement of pastoralists and directly through the techniques of cultivation of peanuts themselves. It illustrates a number of recurring themes in this book, and is summarised by Fig. 7.1.

This study illustrates both a spatial displacement of food crops by cash crops, as well as of people (pastoralists). The latter process has been illustrated by the examples of Zambia, South Africa, Peru and Mexico, but is also widespread where new classes or ethnic groups (or both) gain control of land previously controlled by peasants or pastoralists. Sometimes it is an emergent class of small capitalists which arises from the 'original' users of the land itself. This process of differentiation and class formation has an immense literature deriving

Fig. 7.1 Casual relationships in soil erosion in Niger (from Franke & Chasin 1980 & 1981)

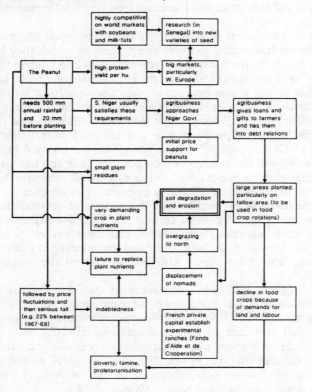

to a considerable extent from the introduction of the Green Revolution into the agriculture of the lesser developed countries (e.g. Ladejinsky 1969; Griffin 1971; Cleaver 1972; Pearson 1964; Pearse 1980; Harriss 1982). In other cases, cash crops can cause a breakdown in other farming operations undertaken by the same farmers. For example, in Sukumuland, Tanzania, the introduction of cotton into foodcrop systems was successful in increasing farmers' cash income. Since there was no outlet for this cash other than in the purchase of cattle their numbers increased greatly, putting great strains on communal grazing areas. Huge stock losses followed and to recoup their losses, farmers increased their area under cotton – further reducing communally-used grazing areas and exacerbating the problem (McCowan, Haaland & de Haan 1979).

In the context of an explanation of soil erosion, we have to examine the implications of people being pushed into marginal places for land-use and land 'misuse'. Spatial marginalisation seldom occurs on its own, of course; it is usually accompanied by other forms of subordination and disruption. For example, Wisner (1976) and also O'Keefe, Wisner and Baird (1977) trace the way in which 'marginals' in Eastern Kenya were pushed into more arid lands. In addition, the build-up of pressure for land in the Kikuyu reserves plus government encouragement of smallholders led to the marginalisation of the weaker sections of the peasantry who found themselves unable to compete with the more powerful and officially backed richer peasants. This led to outmigration to the more arid areas, giving rise to population pressure. Furthermore, the Swynnerton Plan, a central part of government agricultural policy, had encouraged the growing of cotton in the lowlands and this created competition for labour previously allocated to grain and chickpea production (both important food crops). Thus a combination of factors was operative, only one of which was spatial marginalisation.

A summary of the situation in Kenya and Tanzania (Dinham & Hines 1983, chapters 3 & 4) illustrates well the process of spatial marginalisation by a rising agricultural bourgeoisie and by the export crops that they grow, with the result that there is overcrowding on land of low agricultural potential, increasing landlessness and poverty with malnutrition. 'Perhaps even more worrying, given the urgency of the need to increase food production is that... agricultural expansion in recent years has been achieved at the expense of widespread soil erosion, depletion of the nutrient content of the soil, and the destruction of indigenous forests' (ibid.: 93).

Three contrasting case studies illustrate different aspects of this concept in more detail. In the hill tracts of Nepal, the extent of differentiation of the peasantry tends to be less pronounced than on the plains of India, and absolute landlessness affects only about 25 per cent of the rural population. However, differentiation is taking place as a result of extreme population pressure (up to fifteen persons per cultivated hectare in the hills) causing a diminution and fragmentation of holdings, and unequal access to land and capital. The establishment of capitalist agriculture in the hills is very limited (Blaikie, Cameron & Seddon 1980) and there is not a strong and expanding class of 'kulak' farmers causing more marginal families to sell their land through the foreclosing of mortgages as is reported to be more generally the case in India. However, weaker households which do not have access to sufficient private land are forced into a number of desperate survival strategies, one of which is to clear the forest illegally, usually on very steep and as yet uncultivated land, and to plant a speculative crop in the small clearing (often buckwheat or potatoes), in the knowledge that the soil will probably be washed away within a season or two. These cultivators (called in Nepali *sukumbhasi*) sometimes already own one or two plots of land,

and the process of spatial marginalisation refers to only part of their production. Thus the 'geographical scale of resolution' is a matter of hundreds of metres, and does not imply total migration and/or the ensuing destructive use of land over many thousands of kilometres.

Another case study which carefully follows through the connections between class, land-use and soil erosion concerns a tribal people, the Sora, in the state of Orissa, India (Vitebsky 1980). The focus here is on a 'rapid and destructive ecological change'. Up to thirty years ago the hill area inhabited by the Sora was covered by dense jungle, while today the hillsides are near-deserts of raw red soil. Shifting agriculture is practised with only three to four year fallow periods, as opposed to over ten years some generations ago. The population has grown only slowly, and certainly at a much slower rate than the rapidly increased destruction of the environment might immediately suggest, so a simple demographic explanation must be rejected. The explanation hinges upon the fact that, as Vitebsky puts it, 'in spite of apparent surplus, the Sora are rapidly destroying their soil by overcropping', and the implication is that the surplus is going elsewhere. There are numerous informal and para-governmental networks of revenue collection on a scale exceeding that of 'legitimate' demands. There are police fines, and local fines and debts ('officials have an attitude of plunder reminiscent of that of Roman provincial governors', p.10). The revenue inspector collects rent for freehold arable (paddy) land, while the hill slopes (used for shifting cultivation) are not subject to tax but are 'owned' by the government for forestry activities and subject to the threat of being made out of bounds to shifting cultivators. In addition, the Sora are increasingly indebted to caste Hindus living in the plains, and a large amount of grain produced by them is eaten in the plains, either sold cheaply in the markets (as the Sora are usually in contractually inferior positions to grain dealers) or used to discharge debts.

There are inflows to the Sora by way of subsidies on bridges, roads, schools, ploughs, and cattle, but corruption, neglect and diversion of funds for personal political ends prevent these investments from having any real effect. Manufactured goods are imported and local skills (pottery, iron-smelting, basket-making) are lost. In sum, the Sora are passing from a state of relative 'techno-self-sufficiency' to a form of integration based on 'techno-dependence'. Within the Sora too, increasingly contradictory relations of production are becoming apparent, between the rich Sora who own irrigated land, and shifting cultivators who own no land. The latter are oppressed by debt clearance to the richer landlords and are becoming increasingly proletarianised.

Finally, a case study of Botswana by Cliffe and Moorsom (1979) traces the colonial impact on pre-capitalist class relations in terms of the privatisation of land and the necessity for livestock to provide draught power for agriculture, both of which have brought about differentiation among the peasantry. Cattle ranchers and rich or middle-income peasants who had in excess of a minimum threshold of land ownership *and* cattle, and had access to a plough (in the case of peasants), were able to accumulate and expand the scale of their production. Poor peasants had neither and the men were forced to seek employment in South Africa and become migrant labour, leaving the women at home, often destitute. Deep privately-owned boreholes were also introduced, as opposed to shallow, hand-dug public wells, which were previously used. Ecological collapse is occurring now because the new boreholes are fixed in location and fewer in number than the hand-dug ones, so that the areas surrounding them are seriously overgrazed. Also the plough has greatly increased productivity, causing over-cultivation in a very fragile environment, particularly around villages. A minority of richer cultivators who have access to waterholes and bullocks can also manage some winter ploughing, which augments their rate of capital accumulation and furthers the process of differentiation – and a decline in soil fertility.

Government policy, mostly contained within its Tribal Grazing Lands Policy, has developed and extended individual land rights and communal ranges with a limited number of cattle on the pastures. There are high-entry qualifications to communal ranges since the policy was formulated predominantly in the interests of large ranch and stock owners. There has been a considerable restriction of the right of access of smaller ranchers and cultivators to pastures, although there are some proposals for group access for smaller cattle owners (which in the view of Cliffe and Moorsom (1979) are not more than an ineffective political sop). In these ways conservation initiated in the interests of large operators has hastened differentiation, and has tended to reserve the environment against the majority for the increasingly exclusive use of the minority, which has led to serious environmental degradation.

5. Spatial marginalisation, private property and the commons

There is an important element in spatial marginalisation and soil erosion which concerns the existence and extent of private property. Often there is both privately owned land and livestock, and also common pasture and forest. Many case studies have attributed deterioration of pasture to the difficulty of regulating privately owned stock or the collection of forest litter and firewood for private use from 'commons' or the 'public economy'. Of course, the commons themselves, or at least land held by customary tenure, are frequently privatised and legally backed by the notion of private property. This leaves less public land which tends to become over-used, and a vicious circle of increasingly desperate and intensive land-use occurs simply because there is no other source of vital resources (grazing, forest litter, timber, and fuelwood). Furthermore, the location of common land is often on steep-sloped and/or marginal areas, where it represents a residual area after private property has been established in the more fertile, less steep lands. Thus it is often the very steep slopes (and the most susceptible to soil erosion) and the watersheds which remain common property and susceptible to the pressures described. Hardin's celebrated and highly influential paper 'The Tragedy of the Commons' (1972) has excited much comment since it was first published. Certainly the mechanism by which each herdsman derives a positive utility of one (or almost one) with each additional animal he grazes on common pasture, but only a fraction of the negative utility of one, is felt by the individual herdsman in a case of overgrazing. Adding together the component partial utilities, a rational herdsman adds one more animal, and one more...leading to the 'tragedy' (in the sense of the 'solemnity of the remorseless working of things') of the commons (ibid.: 137). In two later works (Hardin 1977; Hardin & Baden 1977), this theme is expanded, and makes political assumptions much more explicit. These are discussed in Chapter 9.3. At this point, it is necessary only to point out that these assumptions are very different, indeed opposed to, those made here. However, the process of the tragedy of the commons is one which both Hardin and this author acknowledge. In this book it is conditions of inequality that are emphasised and are considered an essential driving force in the over-use of common land. This inequality is often underpinned by encroachment of the privatisation of land, which forces marginalised people to use the commons more intensively and contributes to the shrinking potential of the commons themselves. It is these conditions of impoverishment which make regulation from above very difficult, as was shown historically during the Enclosures in Britain (Roberts 1979), and in the Sahel today by Thomson (1977).

   In many parts of lesser developed countries a clear distinction between community-controlled lands (often pasture, bush-fallow or forest) and privately used or owned lands (usually agricultural land for crops) in terms of their control by persons of different economic power, is simplistic. There are often complementarities in the production process, social relations of production and in the movement of soil nutrients between different areas, different land-users and different groups of people. Environmental deterioraion in community-controlled lands has implications for production in more intensively used and privately controlled lands. Also there are common processes of the 'underdevelopment' of agriculture of an entire production system (i.e. for all types of land-use and people).

One of the most common types of movement of nutrients is the transference of fertility from one system of land-use to another. This may occur as a result of cattle or small stock being fed crop residues from one area (often an unirrigated area), but their manure being used in another (perhaps, though not necessarily irrigated) area. Sometimes the process occurs over quite small areas (e.g. Dumont 1970, Blaikie 1970, 1971) and may bring about soil degradation in the outlying areas. Maps of varying fallow periods within Indian villages in the semi-arid zone attest to this, and aeolian erosion of outlying areas with a high percentage of fallow is one implication (Blaikie, 1970). Complex complimentarities between pastoral and agricultural systems are essential in many semi-arid areas (McCowan, Haaland & de Haan 1979), and disruption of one system or the flow between the two has immediate deleterious effects on both.

A more complex example is afforded by Harriss (1982) for a village in North Arcot District, Tamil Nadu. Figure 7.2 explains his major hypotheses concerning the forces of production in the village study.

Harriss identifies the structural determinant of underdevelop-

Fig. 7.2 Environmental decline in North Arcot, India (from Harriss 1982)

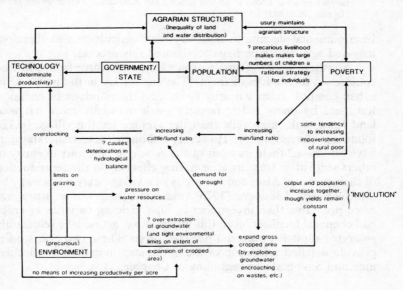

ment of the entire village agricultural system as an inequality in access to land and water, maintained by usury, which leads to increasing poverty of the majority of the population. Rapid population growth, itself a result of poverty, reinforces inequality and also leads to increasing person/land and cattle/land ratios. In this area cattle are absolutely essential for ploughing and the production of milk, and increasing numbers are therefore not attributed to any special Hindu religious principles. Cattle are grazed predominantly on communally-owned pastures on the interfluves, and their increase in numbers has led to a pressure on water resources, overgrazing and erosion. It has also probably led to the siltation of tanks used for irrigation and a reduced and more unreliable supply of water. Population growth has also led to a reduction in publicly-owned pastures through the expansion of privately cropped areas with the aid of wells and pumps for those richer farmers who can afford them – thus further increasing livestock/pasture ratios on the communal land. What is not clear from the analysis (and it is difficult to see how the author could have obtained the necessary data) is whether there has been a secular decline in say, the last fifty years in the application of organic manure. In any case, he cites evidence for yields being more or less stationary in this period.

Until the introduction of new technology, particularly improved and high-yielding varieties of paddy and chemical fertiliser:

> Both in technological and investment terms, agricultural productivity per acre had nearly reached the limits of what was possible without industrial-scientific inputs, and the increase of population had therefore steadily reduced the surplus product above what was needed for subsistence. (Elvin 1973: 312 in Harriss 1982: 103)

Environmental decline in both dry land agriculture and therefore irrigated agriculture perhaps accelerated this process.

So where were the missing soil nutrients to come from and who could avail themselves of them? The answer lay in the purchase of urban compost from a nearby town, and the purchase of chemical fertilisers by many richer farmers (with markedly more irrigated land from tanks and wells than the average for the village (p.126 following Harriss 1982). However, although these richer farmers have purchased their way out of the eco-crisis, the vast majority of others still suffer from its accelerating effect upon the reproduction of their poverty. Also, another study of a village sixty miles away by Djurfeldt and Lindberg (1975), indicated that, where usury was more profitable than investment in high yielding varieties of crops and chemical fertiliser, even this 'progressive' action by a reluctantly emerging capitalist class did not take place. These two case studies provide detailed refutation of the optimistic quotations by Beckerman and Simon at the beginning of Chapter 2.1.

6. Steep slopes, history and soil erosion

So far the discussion has focused upon individual land-users, their economic and social relations with others and with the environment, and it was noted that there was an enormous variety of ways in which peasantries related to the world economic system – from virtually independent subsistence farming, paying few taxes and purchasing a tiny proportion of the household's needs, to situations where the peasant has some rights to cultivate no more than a tiny parcel of land and relies mostly upon wage labour. It is argued in this section that this variability can partly be accounted for by a particular history of steep-sloped areas in which these land-users live, which in many respects have a homogeneity and differ from other 'flat' areas. It will be argued that many steep-sloped areas today suffer from a common syndrome of environmental deterioration, demographic pressure, political and economic subordination and the partial preservation of an ancient or feudal social and economic structure. The following hypothesis about the historical development of steep sloped areas in lesser developed countries draws attention to areas which were *not* colonised, nor were comprehensively under colonial control or administration although they were deeply affected from without.

Steep slopes usually increase soil erosion, relative to less steep slopes. The positive correlation between slope angle and soil loss is well-documented (e.g. Young 1960: 120, and 1976: 1; Hudson 1971: 183; Morgan 1979: 2) and therefore, *ceteris paribus*, a given land-use will produce greater soil losses on steeper sloped land. Generally speaking, this constraint to sustained crop yields has led historically to a number of different outcomes. In some cases, heavy population growth has induced agricultural innovations (along the lines suggested in Boserup's two books, 1965 and 1981) including terracing and labour-intensive crops, particularly paddy. In other cases, agriculture has remained fairly extensive, often involving forms of shifting cultivation and the keeping of livestock, while in others agriculture and pastoralism form only part of the people's livelihood and trading (and raiding) are very common supporting activities (e.g. the Berber tribes of the mountains of the Mahgreb, the Shan and other tribes of Burma and Thailand, and the Tibetans). These steep-sloped areas tended not to provide a large agricultural surplus on which state formation could proceed – in spite of the fact that in semi-arid areas they had often been settled since ancient times, providing the opportunity for small-scale irrigated agriculture. Unlike in many pre-capitalist states whose heartlands were often located in areas of agricultural surplus extracted under feudal or slave relations of production, agriculture in steep-sloped areas remained predominantly for subsistence and not for sale, but sometimes involved limited barter and trade. The dominant political organisation tended to be tribes, loose confederations of tribes, and petty chiefdoms.

With such a fragile and limited agricultural base, trading and raiding (often on neighbouring plains which offered richer pickings) brought the inhabitants of these areas into contact, often in the form of skirmishes or war, with proto-states nearby. The ruling classes of these states maintained a feudal structure in which surpluses were appropriated from land-users (usually peasants) by a military aristocracy and passed on to the emperor or ruler, who used this surplus to buy arms, the allegiance of key members of the aristocracy, to indulge in conspicuous consumption, build palaces, walled cities or devotional buildings, and to maintain a standing army. The latter was used to try and expand territory in order to control more surpluses up to the extensive margin of the strategic and political control of the state. Thus taxes, tribute, labour (for public works, and building and to serve in the army) and sometimes raw materials were frequently sought by rulers of states of hilly areas and their inhabitants. In addition, many pre-capitalist kingdoms relied on long-distance trade (e.g. Hopkins 1973; Terray 1974 for west Africa; Lacoste 1974 for north Africa, the latter illustrating well the conflict between the hill people and the armies of lowland states which constantly tried to control the gold trade and guard the opportunities to amass enormous wealth from marauding hill tribes). Depending upon the relative military power of the two sides, sometimes the proto-state levied taxes, had access to raw materials (frequently forest products) and generally contained the military threat. Sometimes the hill people were able to extend control into the plains for a while (e.g. Prithvi Narayan Shah during the unification of Nepal in the eighteenth century at the expense of the decaying Muslim kingdoms of the Gangetic Plain).

A fundamental shift in this state of affairs occurred when Europeans arrived both in the period of the old mercantile imperialism prior to 1850, and also in that of the new imperialism with more ambitious aims after 1850 (Cohen 1973). While the major source of surpluses (labour to work the mines or plantations, land for plantations or big farms, and land revenue) was in the plains, the hill areas remained largely peripheral to imperial economic interests. They frequently remained a troublesome frontier in the era of the new imperialism and many imperial wars concerned the pacification of hill peoples, although complete annexation of their territories was seldom considered worthwhile. In some cases, too, a buffer between empires provided additional advantages to the imperial powers. Many hill people were skilled fighters, knew the terrain in which they were fighting, and could melt into the mountain fastness and forests as soon as the imperial column threatened. The Afghan wars *war.*

of 1839-42 and 1878-80, the war between the East India Company and Nepal, concluded by the Treaty of Sugouli in 1823, those against the hill tribes of what is now Burma (three during the nineteenth century, ending in 1886) are some examples. Another apposite case study is that of the Moroccan Rif and the extraordinary exploits of Abd al Krim (Hart 1976), and the subsequent transformation of the peasantry after the pacification of the region (Seddon 1981: 141–64). All the attributes of the historical unfolding of steep-sloped areas are there – commerce and piracy plus pastoralism and a fragile agriculture; a long history of war with more powerful lowland states; heroic resistance against two imperial powers (French and Spanish); eventual defeat; establishment of a pattern of outmigration to foreign armies and later to temporary employment in large-scale enterprises far away; differentiation of the peasantry at home; and serious environmental decline. Many of these struggles ended, however, in military stalemate and treaties or understandings were devised between the imperial power and local leaders. Varying degrees of political and legal independence were retained, but with treaty clauses which allowed imperial economic interests to operate as best they could, given that occupation and administration was either an unreachable military objective or simply was not worth it. Those interests were, of course, varied, depending upon local circumstances, but frequently implied the recruitment of labour to imperial armies (to police other parts of empire), and to mines; to have the right of access to markets within the hilly areas where there was consumer potential, and to raw materials such as precious metals and timber (e.g. for sleepers during periods of railway construction; Schlich 1889 about north India).

In such negotiations the local ruling class with whom the imperial power dealt could often strengthen its internal position and benefit from the new accommodation (by acting as contracting agents for the export of raw materials and cornering the franchises on imported goods). Thus these steep sloped areas often managed to retain a degree of internal autonomy, and the relations with capitalism tended to be mediated by the interests of a local ruling class:

> The preservation of these antique, semi-feudal or patrimonial states as the administrative and military substructure of imperialist exploitation at the same time discredited the old ruling order and provided it with a means of survival. It was precisely the mortgaging of the national economy and resources to foreigners (treaties, concessions, abandonment of tariffs) which supported the regime and created a ramified network of compradors, contractors, profiteers, and complicit officials in whose interest the system worked. (Rey 1959: 70)

The implications of this pattern of historical development in hilly areas were:

(a)   Trading and raiding were severely curtailed. Trade was controlled by imperial powers or loyal middlemen, and frequently the goods previously traded ceased to be important once the native aristocracy had been destroyed or assimilated by the imperial power. Hence, products such as civet, ivory, medicinal herbs, and other luxuries produced or traded by hill people declined in importance. This put pressure on agriculture as the main source of income.

(b)   These areas were sometimes used as a reservoir of labour to be employed on large plantations or farms or distant mines. The peasant farm was used to cross-subsidise capitalist enterprises outside the region in the manner described in sections 2, 3 and 4 of this chapter. Agriculture, then, was not forced to provide most of the livelihoods of the population. As population grew, remittances often came to 'prop up' the agricultural economy.

(c)   These areas did not exhibit the marked differentiation of the peasantry which occurred in many agriculturally productive areas under imperial rule. The peasantry was left more or less intact and not dispossessed of its land. Frequently customary forms of land tenure were not replaced by notions of private property.

(d)   Population pressure due to a falling death rate (thanks to both rudimentary preventative health measures, improved communications, famine relief, and political stability), a more slowly falling birth rate and a stagnant productive base, started to be felt. The result was a diminution and often fragmentation of holdings, extension of the cultivated area and destruction of forests for agriculture and as a supply of fuel. Population pressure in some instances can also be felt with *declining* population where agricultural technologies have regressed because of the impact of colonisers (e.g. the apocalyptic impact of the Spanish upon Indian populations in Central America; Boserup 1965: 42).

(e)   All these factors accelerated the loss of the ability of inhabitants to be self-provisioning. Landholdings were inadequate, agricultural technology not suited to intensive production, and a wide range of needs could not be met internally (household utensils, shoes, clothing, lighting, building materials, and agricultural implements). These were imported increasingly and paid for by exporting labour (see (b) above) and petty commodity production for the market – which further increased demands for land and the forest.

Each of these implications is linked to the others, and they are mutually reinforcing. Environmental degradation ensures, exacerbates and accelerates the syndrome in the ways outlined as feedback effects of soil erosion in the scheme presented in Chapter 6. In the post-colonial period, most of these symptoms still exist since the structures which brought them about remain intact, and even the revolutions that have occurred in Ethiopia and Afghanistan do not imply a rapid reversal. These areas continue to suffer from a lack of public sector assistance (if they form a part of a state), and economic and political peripheralisation.

The applicability of this sort of generalisation is a problem. The presentation of a generalised 'archetype' is admittedly analytically weak, and there are always exceptions to particular aspects of the archetype, such that it applies nowhere when strictly examined, and increasingly everywhere as its distinguishing characteristics are progressively diluted. However, it is claimed that there are examples enough for this generalisation to have some validity – Afghanistan and the North-West Frontier region, Nepal, Thailand, Ethiopia, Lesotho, Kabilya in Algeria and the Rif Mountains in the Mahgreb in general, the remoter parts of Andean Peru, Colombia and Bolivia, Albania and parts of central Yugoslavia.

The implication for understanding soil erosion is that in such places there may be a set of problems of a long historical standing, and that *external* political economic relations are intimately connected with soil erosion. Thus recession in Europe will affect the flow of *gastarbeiter* from Europe's eroded mountainous periphery, remittances to the regional hill economies, investment in agriculture and decision-making in land use. The rate of soil erosion in Lesotho is intimately tied up with the Republic of South Africa's recruitment policy for the mines. It has been said informally to the author by Indian military personnel that India's continuation to recruit Gurkha soldiers from Nepal is as much a strategic decision as a military one in that a continuing flow of army pensions and remittances serves to prop up the crumbling hill economy in a politically sensitive area. The prognosis for these areas or states is often bleak, and the degrees of flexibility in political choice and production opportunites in general are often very limited.

Chapter 8

The other side of the coin

1. Introduction

The previous two chapters have focused upon the problems faced by small producers in meeting the costs of their subsistence, and their relationships with others, the state and the environment. One of the major conclusions was that small producers cause soil erosion because they are poor and desperate, and in turn soil erosion exacerbates that condition. A set of socio-economic relations, collectively called underdevelopment is at the centre of this poverty. Attention in this chapter shifts to large enterprises and *their* relations with others, the state and the environment. We have already discussed their relations with these small producers, and how large enterprises may thereby indirectly cause soil erosion, although paradoxically, in some cases, carry out effective conservation on the land that they themselves use. However there are many circumstances in which this is not the case. Logging concessionaires extract timber from tropical hardwood forests in ways that are highly detrimental to land and water conservation, and at rates which endanger the future livelihood of many people. Both foreign and indigenous enterprises clear forest and use the land for ranching with serious environmental consequences, and if they are allowed to, cultivate land for short periods, extract the stored-up fertility of the soil, and then move on. The technology they use, often supported by national research stations and those with the opportunities for selling the technology, is sometimes ill-suited to the tropics. Thus, large-scale producers can *directly* contribute to soil erosion. It is likely, however, that the actual area used by large enterprises is relatively small compared with that used by peasants and pastoralists. The actual area of land owned or contracted to transnational companies in Africa, for example, is quite small, but it is extremely difficult, if not impossible, to get data (Dinham & Hines 1983: 28). In any case there are a great many different ways in which transnationals control land-use other than holding privately owned plantations or ranches. In some areas of lesser developed countries notably in central America and in Amazonia, both in absolute and relative terms, the areas held by transnational companies are much larger. In almost all cases, large enterprises, often aided by international finance in the form of a 'project', are located on the most fertile land. It seems likely therefore that, on balance, the contribution of large enterprises to soil erosion is indirect (through the processes described in Ch. 7) rather than direct. However, there are many instances of a direct contribution by large enterprises by their own use of land, and it is to this issue that we now turn.

2. Large enterprises using land

One of the important ways in which large enterprises use land in lesser developed countries is in obtaining concessions for the extraction of timber. It has long been recognised that in many parts of the world, particularly South-east Asia, Melanesia and parts of tropical Central and South America, forests have been degraded or even depleted (cleared) in an irrational and wasteful fashion. There is also a debate about who is to blame in which the logging concessionaires and the shifting cultivators are the two principal candidates. First of all, the actions of the two candidates are different sides of the same coin. This comes about through spatial marginalisation and of surplus extraction from poor shifting cultivators. Where there is less competition for the use of the forest, there are cases where it is the indigenous population which is primarily responsible for cutting down the forest as a result of rapid population growth. However, most of the difference in perspective of various commentators probably derives as much from political standpoint as from the local circumstances themselves at the time. For example, compare the difference in emphasis of accounts by Plumwood and Routley (1982), with that of Fontaine (1981), Spears (1982) or Lanly (1982). There is a broad distinction between open-tree formations where agriculture and fuelwood demands are more responsible for degradation and depletion and closed-tree formations where logging is the main agent (Lanly 1982: 77f.).

The extent of deforestation has only recently been revised upwards from NASA satellite pictures. Previously statistics were compiled from national government estimates; and these have proved to be wildly inaccurate (e.g. in India, CES 1982). In the Philippines, government sources claimed that 57 per cent of the land area was under forests, while satellite pictures gave the percentage as 38 per cent (Grainger 1980: 6). Even national statistics indicate that deforestation in some very large areas is alarmingly high. For example, in tropical America, the 600m. ha Amazon rain forest is being depleted at a rate of up to 11m. ha per year (Myers 1979), of which 38 per cent is due to large-scale agricultural schemes and large-scale ranching and 31% to small farmer occupation (see also

White 1978 for Peru). In central America, the tropical rain forests and pine forests are threatened in many places and annual depletion rates of 3.5 per cent for Costa Rica and El Salvador are occurring. In Asia, the circumstances of deforestation are highly variable. In South-east Asia, particularly Malaysia, Thailand, Indonesia, Papua, and the Philippines, deforestation rates have been exceedingly high and large logging concessions have had a major responsibility. For example, in Thailand, the total area under forest has declined from 63 per cent in 1947 to under 25 per cent in 1977, with a present annual deforestation rate of 3.6 per cent. Most of the devastation has occurred in the north-east where both shifting cultivation and to a much lesser extent logging companies have contributed to the problem. The impact of deforestation has been serious, e.g. the siltation of the Bhurnipol hydroelectric reservoir (Kilakuldilok 1981; Brow 1976; Khanbanonda 1972). All the major forest areas of the Fiji archipelago and Solomon Islands have been allocated to foreign concessions, while the Madang project in Papua New Guinea is reported to have had disastrous environmental consequences (Routley & Routley 1980). The logging policy of the Indonesian government has particularly invited criticism (Plumwood & Routley 1982; Grainger 1980):

> It seems entirely unlikely that Irian Jaya will escape the Indonesian Government's policy of massive forest exploitation of the outer islands, and the simple and effective idea of environmental destruction as a tool for social change, or that the Melanesian customary land tenure system will be allowed to stand in its way. (Routley & Routley 1980: 60)

It is interesting to note that Lanly (1982) in his detailed analysis of tropical forest resources of some 100 pages, hardly once mentions the role of logging concessionaires in forest depletion or in environmental deterioration. Shifting cultivation appears frequently as a major cause. In the case of Indonesia, it could be hypothesised that there are special economic circumstances which make it all the more likely that Routley and Routley's dire predictions may fulfil themselves. Indonesia has embarked upon a costly development programme paid for by oil reserves. With current uncertainty about the price of oil and a marked fall during 1983, the necessary foreign exchange to finance the programme (and the expensive tastes of ruling groups) will have to be found increasingly from the export of timber.

In Africa, there are large areas of rain forest still left, but in western Africa the annual rate of deforestation may be as much as 5 per cent, and in Nigeria and Ivory Coast long-established commercial logging is causing the forest to be rapidly depleted (the forests of the Ivory Coast are projected to disappear by 1985, Roche 1978). The open forests of the savannah belts of Africa have also suffered as a result of overgrazing and fuelwood demands, particularly in the Sahel which is discussed elsewhere in this book.

These summary illustrations do not imply that forest reserves should not be exploited. They do indicate that the rate of exploitation will lead to complete destruction of the forest in many areas often within twenty years, and it will be suggested that, in the case of logging concessions, timber extraction is carried out in such a way as to minimise the economic benefit either to the host country as a whole or to people living in or near the exploited forests, which in many instances causes very serious environmental degradation.

It has been widely claimed by both members of national governments as well as other commentators that transnational companies are in a very powerful position in negotiations with national governments enabling them to exploit the forests in ways that suit them rather than the long-term interests of both the nation or local people involved. Specifically, these companies seek to reduce the economic benefits from logging to host countries by neglecting or even blocking forward linkages such as saw mills, plywood or chipboard manufacturing. Second, they tend to pay very low prices for the timber by a variety of honest and dishonest ruses, among them being the use of superior local knowledge of the concessionary area, threats to move operations outside the country altogether, and other unofficial inducements to negotiating officials. Third, they attempt to maximise repatriation of profits, and to avoid contributing to the cost of either reafforestation or training local personnel. Fourth, they use the transfer pricing system to minimise royalties or taxes to national governments. Fifth they 'cream' forests in a manner that would never be allowed in developed countries (Leslie 1980). Sixth, once the negotiations have taken place, the implementation of specific conditions of the contract is frequently not carried out.

The reasons why these claims may have widespread substance are that transnational companies often have superior technical knowledge both of the area involved and of the logging industry. Their negotiators have very large personal stakes in the matter – it pays them to be well informed and negotiate in the strongest possible manner to maximise profits, to retain maximum control to do this and maintain it in the future. They face government bureaucrats who do not have direct personal gain (or loss) as an extra motivation, whose system of administration may be 'imperfectly open' (Leslie 1980) in that inconsistencies in government policies and incomplete information face the government negotiator. Incomplete information upon the actual volume of timber of various types, upon the probable costs to local inhabitants and to others in the same watersheds through reduced production as a result of flooding and siltation, and the inability to coordinate the negotiator's position to an overall environmental policy, all serve to reduce the strength of his position.

In the same way as in the nature and extent of the impact of soil conservation policies, the physical outcome of negotiations over logging concessions is a matter of political struggle. The adverse effects of indiscriminate logging, especially clear-cutting for the wood-chip industry, inordinate damage to other trees in selective felling, the loss of livelihood suffered by local forest-dwellers, siltation and flooding and so on are not experienced by the government negotiators themselves nor the ruling classes they represent. Indeed, 'successful' negotiations ensure some foreign exchange, which may ease balance of payments difficulties and may offer rich pickings for individuals in the country involved (labour and transport contractors, engineers, land owners where the forest does not belong to the state) as well as other companies indirectly which become involved in supplying other necessary infrastructure, particularly roads, construction equipment and bridges. In addition:

> Internal inequality and resistance, and the repressive, militaristic character of the national governments concerned, creates an obsession with strategic and 'national security' considerations (which favour the removal of large natural areas which could serve as a base for organised resistance), and also increase the emphasis on the exploitation and subjugation of nature. (Plumwood & Routley 1982: 11)

Certainly military and strategic considerations are important (as the Turks found when confronted with Greek, Albanian and other Balkan armed resistance in the forests of the area, resulting in large-scale destruction of the forest), and often reinforce economic ones. This assertion, however, is difficult to 'prove' since it is not one that is publicly spoken about or is developed explicitly on paper. This type of explanation therefore does not set much store by the effectiveness of the improvement of legislation in logging (e.g. Schmithüsen 1979) and 'strengthening' institutions involved in forestry (e.g. the departments at central and local level, local surveillance, monitoring systems of forest use, research and development training, etc. Spears 1982). While all these changes must be in the right direction, and certainly can do no harm, the underlying political-economic relationships concerning large-scale logging undermine these attempts.

Turning now to large land-users in agriculture and ranching, it is both their relationship with the state and the technology which they use which sometimes encourages them to mine the soil. With regard to the agricultural technology most favoured by large enterprises, it is a matter of technology transfer from the temperate to tropical conditions as well as the relative factor endowments of the entrepreneur which largely determine the way in which agricultural land is used. Large scale and usually capital-intensive land-use to produce a crop for an international market usually implies the necessity for enhanced control over all aspects of the growing environment. In the case of annual crops, the product itself has to be harvested in a timely and accurately plannable manner. Quality control, uniformity, and a high degree of specificity of desirable characteristics in the product all make for single stand planting, a high degree of mechanisation in tillage practices and harvesting, control of plant populations and line planting to aid mechanisation and maximise yields, selective breeding of cultivars, and an intensive usage of chemical fertilisers, pesticides, fungicides and weedicides. Also large-scale production of a single crop tends to discourage long crop rotations, or sometimes an absence of rotations altogether. Problems sometimes arise because this technology does not suit tropical conditions. Compaction of the soil by the use of heavy machinery, deep ploughing causing moisture loss and exposure of humus to the sun and to leaching by heavy rainfall, the build-up of salts or toxicity in the soil due to irrigation without adequate drainage, physical soil removal because of low plant densities, line planting and low foliage protection (leaves are often bred out of cultivars in the pursuit of higher yields), are some of the problems that are caused by this technology.

3. Accumulation and degradation

In Chapter 5.4 it was suggested that those classes and groups adversely affected by soil erosion were often small rural producers and pastoralists, the working class, and the disadvantaged (old, poor, sick, unemployed or semi-employed) in the towns. In general the most powerful classes were those whose members were least affected or could adjust most easily. This section takes up the actions of the most powerful, and discusses their options when faced with the symptoms of environmental degradation, such as falling crop yields, natural fuel shortages, water shortages, floods and siltation.

Two hypotheses are suggested. The first states that only when soil erosion is perceived to affect adversely the process of accumulation will it bring about attempts to reduce it. The second hypothesis states that this does not happen often. Let us examine the first hypothesis.

There are two elements in this hypothesis – the economic circumstances of those who are accumulating, and secondly their possible course of action when they perceive that soil erosion could seriously affect profits. In the case of transnational companies, mining the soil and then moving on to other regions or other nations altogether is quite common. There are many examples of inappropriate intensive cultivation of export crops leading to soil erosion, followed by abandonment. The case of Bud Antle Inc., a large California-based food conglomerate in Senegal is given by Dinham and Hines (1983: 31-3). Quality vegetables were grown in Senegal and flown off to markets in Europe. By 1976, the project had virtually failed and the Senegalese government (which had formed a joint enterprise with Bud Antle Inc.) was left with eroded soil and imported machinery which could not be maintained (ibid.: 32). A rice-growing project in the Caramance area (ibid.: 151) is having similar problems. The case of large farmers acquiring cheap credit in Brazil, mining the soil and moving on to catch up with the 'hollow frontier' has already been described.

There are a number of circumstances which, if occurring together, will encourage the owner of capital to use up the natural resource content of the soil and then withdraw from the area altogether. The first is the opportunity to make a very advantageous purchase of cheap land (either outright or a lease, which exacerbates the situation). The second is the ability to negotiate (often with a national government of a lesser developed country) the provision of other infrastructure such as roads, airstrips and port facilities. Both of these reduce the fixed nature of the investment, and many transnational companies have the skill and negotiating power to secure these advantages. The third circumstance is the nature of many tropical soils, which are not suited to long-term intensive agriculture without expensive importation of chemical nutrients. The fourth is the ability to find alternative investment opportunities elsewhere. The fifth, which can often occur, is unsettled political conditions in the country where a transnational company has invested and poor relations exist between the two (where compulsory acquisition of shares by nationals or the state or disagreements over repatriation of profits are frequently the points at issue). Taking all these circumstances together there is a continuum of commitment to the conservation of natural resources from the short-term forest contractor described in the previous section. Then there is the short-lease contract of ten years for the cultivation of a specified area of land. Next there occurs the outright purchase of land and its abandonment whenever it is uneconomic to cultivate without capital investment or greatly increased production costs. Lastly there is the farmer who expects to spend a lifetime of investment and work on the farm. Many transnational companies and some nationals in senior government posts (who can use their position to negotiate cheap loans and the purchase of land) find themselves operating within these five circumstances, and therefore are directly responsible for laying waste large areas of land – as well as displacing subsistence food producers who are forced to overexploit the soil elsewhere. Of course there are transnational companies which do adequately conserve the soil when their investment has been considerable. It is difficult to substantiate the two hypotheses with data other than by example (which is a poor substitute). The pulling-out of companies when the productivity of the soil has been reduced is not an event which is systematically recorded, and there are good reasons why it is not reported at all.

There are other circumstances in which accumulation possibilities may be threatened by soil erosion, and these concern hydroelectric and irrigation schemes where small farmers in the affected watersheds may accelerate siltation of reservoirs and the design of turbines and even the dam structure itself. If this is the case, then it is the interests of the industrial bourgeoisie who will use the electrical power, and other farmers (in the case of a canal irrigation project). Many examples are cited by Grainger in Java (where slash and burn cultivators threaten the irrigation canals in the Solo River catchment), in the Philippines (in the Agno catchment), in India and Pakistan where a large number of dams for power and irrigation have suffered seriously from savage floods, siltation and reduced lives of reservoirs (CES 1982), and in Brazil (particularly for the Tucurni hydro-electric scheme to provide electricity for proposed aluminium smelting operations at Trombetas). Here the political calculus is more complex than in the case of a transnational company because there is no question of transferring capital elsewhere – the dam is by its nature fixed in location and a large fixed investment has to be made. Where there are large numbers of cultivators living in the catchment (e.g. in most of the Himalayan foothills), it may not be politically feasible to force through radical and effective measures. For example, in preliminary reports on the feasibility of the Mangla Dam in Pakistan, it was claimed that up to a third of the population of two million would have to be removed. Frequently it is a problem which is beyond the power of the state. There are also other factors which deter governments from taking politically delicate action. Often it is difficult to attribute accurately the contribution of sediment load to cultivators in the catchment and to assess the impact in removing some of them or persuading them to alter their agricultural or pastoral practices. Also the dam can merely be written off over a shorter period, or other measures such as silt-trapping devices taken. From the point of view of private entrepreneurs, any extra capital cost may not be reflected in the price of electricity which is generated, and therefore they may not feel the need to make their influence felt through government to reduce soil erosion and the rate of deforestation.

The process that is being described in Chapter 7 and here in sections 8.2 and 8.3 is one of incorporation of natural resources and labour power in lesser developed countries into the world economy – a process which was greatly accelerated during the latter half of the nineteenth century, and which continues today. There are many who believe that this is essential in the modernisation of the economy, and is a necessary objective and process in development. Such Marxist writers as Warren (1980) see the process of industrialisation as one which is creating the essential basis for a socialist transition, and others of very different political opinions see the incorporation as progressive, in its own right and as an end to itself. Whatever view the reader takes, the process involves an increasing removal of control, both cultural, economic and political from the local to the national and international. Friedmann and Weaver (1979) identify the dismantling of territorially organised units by their integration and dissolution in the national and international economy. Some readers will argue that this process allows the principle of regional comparative advantage to operate and facilitates the spread of innovation and increased productivity. If this is so, then the problem of widespread environmental degradation which is undeniably the result of this process, has to be faced – and presumably written off as the inevitable result of development, which can be offset in an aggregate sense by technological advances, as Simon (1981) agrees. This accelerating removal of control of soil and water resources has meant that these resources are less and less 'local', but global. These natural resources support local people when utilised in agriculture and pastoralism in the creation of use values. However, they are now being controlled by new classes both directly and indirectly in many different ways. The state or private contractors lay claim to the forest and clear it. Large farmers buy up small farmers' lands or enclose parts of the commons. Small farmers have to grow cash crops, and some improve their incomes by doing so (but many others, as this book attests, do not). Competition between food and cash crops; between fuel, game, wild food and commercial timber; and between the classes who produce for use or exchange, has resulted in a loss of local control of use over these resources. Just as it can be argued that there is less and less justification in referring to 'national capital' (implying that it is now international capital not under the control of national governments, Holland 1980), it could be argued that the concept of a national soil and water resource is increasingly questionable. Therefore, when these resources are under threat through large-scale forest clearance and/or soil erosion, effective political means to conserve them do not lie in local or even national hands.

Chapter 9

What now?

1. Overall prospects for soil conservation

A principal conclusion of this book is that soil erosion in lesser developed countries will not be substantially reduced unless it seriously threatens the accumulation possibilities of the dominant classes. In various combinations and alliances, these classes are the national and international agricultural bourgeoisie, industrial capitalists, and various related groups such as export-import agents, commission agents and government officials themselves. The degree of threat, it is argued, has to be very substantial because the direct impact of accelerated soil erosion is difficult to measure, diffuse in its effects, and is often a gradual process and patchy from place to place. Also it is these classes that can best adjust to soil erosion and feel its effects least, both economically and experientially in day-to-day living. Therefore they are inclined not to be mobilised by the issue of soil erosion.

On the other hand, small-scale land-users often directly cause soil erosion, because they are forced to do so by social relationships involving surplus extraction. Conservation under these circumstances tends to undermine the security of their livelihood in the short and perhaps medium term. They may try to conserve their environment, but cooperation is often difficult because of their individual mode of earning a living (individual survival strategies), and because many of them are under such pressure to secure a livelihood that the short-term costs are prohibitive. The rhetoric is that the state has to intervene, and therefore new institutions are created (e.g. watershed management committees). However, for reasons of conflicting interests, ignorance of local conditions and the over-riding concern of government to increase control over peasants in the name of development, these centrally created institutions tend to fail.

Although an economic explanation to soil conservation policies is the determining one, not all people, nor their ideas, reflect economic imperatives. We have seen that ideas about soil conservation may have considerable independence from a deterministic economic explanation, and certainly are varied. The ideas can be conceived of as a civilising mission on the part of Europeans, of 'less-developed people' who need helping from themselves, a kind of missionary zeal experienced in colonial Africa between 1850–1950. They can be conceived of as a major foreign policy component, along with population programmes and new inputs into agriculture, as a strategy to contain world communism. They can be conceived of as a moral imperative to respect Nature, or on the part of peasants and pastoralists as an integral way of ensuring a continued livelihood. All these ideas find expression in conservation policies from time-to-time, and sometimes contradict those powerful economic interests already discussed, although a broad and long-term view usually indicates that they follow rather than run against them. However, these contradictions exist, and must be recognised and utilised. Otherwise an over-determined, crude and uniformly pessimistic analysis will prevail.

Unfortunately, soil conservation policies do not usually serve powerful economic interests as, for example, land reform movements. In the latter case, land reform in many countries of Latin America and South Asia was a platform for a newly rising agricultural capitalist class to oust backward feudal landlords who blocked their demands. In terms of redistribution of land to small peasants many land reform policies failed, but in terms of the political objectives of their proponents, they substantially succeeded. Therefore the prospects for conservation policies which manage to turn around the rapid depletion of soil and water resources on a world scale seem rather poor. Appeals to reason, to the moral duty to conserve nature or to help the poor peasant and pastoralist through conservation simply have little to offer those who would carry out these policies.

In these circumstances, there appear two ways forward – rhetoric, and what we may call 'deflected action'. Both solutions call for a prodigious output in the form of seminars, conferences, reports and even financial commitments by foreign aid donors (in the same way as described by Franke & Chasin 1980: 145, about the response to the Sahel famine). One particularly ironic example is the FAO sponsored conference on appropriate technology in forestry (FAO 1982a) which took place in India at the same time as the controversy over that country's proposed Forest Bill – the rhetoric in the contributions to the seminar seemed (and did) come from another world from the 'technology' of repression of forest dwellers, massive deforestation, wasteful extraction techniques, brazen revenue maximisation in the short term by state institutions and so on.

The term 'deflected action' refers to peripheral and support action instead of the real business of implementing soil conservation. It includes an array of projects and programmes which *can* be implemented, which are important and often essential social and physical infrastructures to conservation programmes, and which can plausibly be justified (provided that actual conservation programmes really do take place). These 'deflected actions' include training programmes, institution building in soil conservation, mapping and monitoring projects, research projects, sedimentation gauging and other physical experimentation, satellite imagery, and above all, conferences. Thus the prospect for effective soil conservation remains rather poor. By now we can hardly register surprise at the following lament: 'The cures for desertification are well known, but although a United Nations conference held in 1977 produced a Plan of Action to combat desertification, these cures had not been put into effect in much of the world by 1980' (UNEP 1982: 250).

2. Strategic choices

A 'strategic choice' of a soil conservation policy means a choice which is both feasible within an existing political economic context and is in step with a *future* direction of social change which is ideologically acceptable to policy-makers themselves. Three propositions have been made so far:

(i) that soil erosion and conservation arise from fundamental structures in society;

(ii) that it is not possible to offer a critique of the processes of social change from the point of view of soil erosion and conservation alone;

(iii) that *all* approaches to soil erosion and conservation are ideological – they are underpinned by a definite set of assumptions, both normative and empirical, about social change.

From these three propositions, a fourth can be deduced:

(iv) that a view of social change has to be taken a priori to any consideration of soil erosion and conservation, although they may play some part in that position.

This last proposition may be labelled politically biased, to which the author pleads guilty. The problem is that all agricultural scientists, policy-makers, academics, and consultants, are politically biased, including those who claim that their activity in environmental policy-making is neutral (refer back to the arguments in sect. 4.1).

It is not the purpose of this book to outline my own particular views on future directions of social change – some have been made implicit already. In any case, as has been said in the last paragraph of sect. 1.3, the exercise would be unfeasible and unnecessary. What follows are some elements of different views of social change. The purpose of this discussion is not to enter into the very broad-ranging debates about development and social change, but merely to suggest that a soil conservation policy must be consistent with these broader elements. The important point about not trying to make the tail (conservation policy) wag the dog (view on development and social change) bears repeating here.

The elements suggested here are not mutually exclusive, they can coexist – and each of them brings with it new problems. There are four – socialist utopianism, populism, rational policy-making and authoritarianism. This list is not a typology of views on social change, merely a number of strands within views on social change which are relevant to soil erosion and conservation.

(i) The first element is one which this book strenuously tries to avoid – a lapse into general 'socialist' utopianism. This assumes that in a future socialist state there would be less tendency to have conflicts of interest over the use of the environment (for example Robinson 1977). Furthermore the essential cooperation needed to limit livestock, reafforest and keep people and stock from the seedlings, adopt soil conserving practices of husbandry and tillage would be brought into existence painlessly. The hope would be that the levels of surplus extraction from direct producers would tend to fall, giving them the opportunity to invest collectively more of their labour in soil conservation works, to reduce total production levels to prudent soil-conserving levels, and to be able to return fertility to the soil by either/or synthetic imported fertilisers or by composting/ mulching. A useful way of setting out to test these hypotheses is by empirical verification, but the problems of identifying a socialist state give the lie to such a straight-forward venture. Experience in China, Ethiopia and Vietnam seems mixed, with a certain amount of success in China, but it is probably true to say that all these countries' conservation efforts were more successful after their revolutions than before. However this under-specified utopianism does not grapple with the inevitable conflicts of interest in any transition of society, even in the Soviet Union which has claimed to be socialist for more than fifty years. There would still be conflicts between private and collective interests, between local and national priorities over land-use and the allocation of labour, between industry and agriculture, over differentials in pay between skilled and unskilled and so on. Therefore, while a socialist transition may make the necessary cooperation easier between the direct producers themselves and between them and government, those general statements are not particularly helpful.

(ii) A different and more applicable approach which acknowledges existing broad power relationships, advocates that the state treats its peasants and pastoralists better, and 'gets off the peasants' backs', for example:

> In economic policy, the priority must be given to raising food production. This cannot be achieved by state direction of peasant producers, but only by encouraging peasant initiative based on their own experience and improving their own material well-being, and defending their own gains against the demands even of the revolutionary state. (Williams 1976: 152)

A similar view appears in Williams (1981) and by implication is almost populist in tone. It appeals to small producers and is strongly opposed to centralised control by business and international agencies (this latter article criticises the World Bank's policy towards peasants). One of the problems with this view is that peasants and pastoralists do not constitute a nation-state by themselves – they exist in a subordinate relationship to another class which usually controls them. Peasants 'left to themselves' therefore is a notion of development without a state. Also they have *not* been left to themselves and have been profoundly affected by both capitalist penetration and socialist revolution. A populist stance leaves open the question of future agricultural development – its relations of production, its structure of decision-making, its rate of surplus extraction and investment and so on. There are few clues to the transition. In the field of soil conservation, a form of populist writing can be found which claims that the knowledge of the peasant and pastoralist of conservation has been disrupted and scorned by capitalism and the scientists yoked to its cause. While it is undoubtedly true that most peasants know their own environment better than most 'scientists', and that it is worth pointing out the usefulness of their knowledge, it is important not to leave it at that. The implication by default is one of populism. As in the 'socialist' utopian case, a soil conservation policy must be formulated in the context of wider development objectives and indeed of a particular direction of social change.

(iii) The third element is one which exhorts, directs, appeals to the government to intervene in a 'rational' manner. This is the commonest and most coherent approach since it does come up with suggestions about what should be done. In the domain of rural development, there have been many shifts of emphasis, sometimes merely of rhetoric, sometimes of practical policy – from the optimistic 1960s (Pearson 1964),when it was thought that large capital and technology transfers made by massive aid programmes would fuel modernisation and 'take-off', through redistribution with growth, employment creation, and basic needs. Each of these phases has been marked by a definite set of policy measures which governments were supposed to carry out. By and large, soil conservation policies were not sensitive to these changes in rural development policy. One exception may be noted in which agri-silviculture and social forestry have been emphasised where both rhetoric and some determined attempts have been made to increase employment and distribute the benefits of these schemes to a wider spectrum of rural producers. A very telling liberal critique of this view is given by Warwick (1982: 39) in relation to family planning programmes. It is essentially a top-down perspective in which policies, once harnessed to the structures of government, can be implemented through the force of top-down authority. This is what is called the 'machine model of implementation'.

> The guiding notion is that the foremost requisites for successful execution of a sound policy is an efficient administrative apparatus...and clear directives from above...Typically, national administrators issue orders fully believing that the weight of their authority, bolstered by effective monitoring and control systems, would be enough to bring about implementation.

More radical critiques have also been developed throughout all these different phases of rural development styles. Their main thrust has been to attack the view of the state as an impartial arbiter, and see the institutions of state as well as virtually everyone in lesser developed countries being drawn into the process of world capitalism. Some of the leading authors include Frank (1967, 1980); Wallerstein (1974); Amin (1974, 1977); George (1976); Warren (1980), de Janvry (1981). The approach to soil erosion and conservation taken here is within this broad critique. Its emphasises relations of surplus extraction as a cause of erosion, and a conflict model of the state, where policies are selectively formulated and implemented (or allowed to languish) according to the interests, balance of power, and tactics of competing classes and groups within the institutions of state. The implications of this approach for what can be done are discussed in the next section, entitled 'practical pessimism'. At this stage it is sufficient to state the obvious – that policy-making in conservation may be the only activity possible for politicians, bureaucrats, academics and foreign consultants, and the only way that any practical headway can be made at all from within government (both central and local). The critique can only alert such individuals to the limitations of working within the state, the compromises it may involve, and help them uncover the underlying assumptions about development and social change in policies, official statements, reports, private conversations of government officials, and in the actual impact of government action in the field.

(iv) The fourth element is authoritarianism. The political circumstances of an authoritarian government must be distinguished from authoritarian policies, although in many cases the policies are of course enacted by the government. An authoritarian government in post-colonial times may come about as a result of seizure of power in a time of national crisis–a threat to national security, a breakdown in confidence in the efficacy of the state (due to 'meddling politicians', corruption, or a breakdown in law and order). A new government is formed frequently after a military coup in the name of national unity, which thereby lays claim to the right to stifle dissent (i.e. dis-unity). For a variety of views upon the politics of authoritarianism, see Finer (1976). The context for authoritarian soil conservation policies frequently arises from different premises and forms of state altogether. As we have seen in colonial African policies, there was a strong element of coercion which derived from an ideology of European superiority and from the necessity to coerce Africans into performing the economic roles required by imperialism during its different phases. The continuation of coercive soil conservation policies derives from the ideological pattern of the colonial model, as well as the continuation of some of the same economic patterns described in sections 2.2, 4.2, and 7.1–7.6. Also present-day authoritarian states may take on an 'êtatist' role in protecting the environment (e.g. South Korea).

Coercive policies may also derive from a rational argument given by such authors as Hardin (1977) and Hardin & Baden (1977). These authors advocate very stern measures to eliminate common land (i.e. to privatise it) and/or to administer it in the name of the state. By so doing, it is argued, it must be acknowledged that everyone does not have equal rights to land. Crowe (in Hardin & Baden ibid., 1977) makes a number of criticisms, two of which are particularly relevant to soil conservation. The first is that the state does not have the monopoly of coercive force, and active resistance by even a few determined people defeats the purpose of the state – a most cogent point which colonial administrators of conservation policies throughout the world have learnt to their cost. Secondly, Crowe exposes the myth of the 'administrators of the commons' as a rational and apolitical body who regulate the commons in the name of the state – 'who shall watch the watchers?' The concept of the neutral state is being attacked here, although different terminology is used.

The only circumstances under which authoritarian soil conservation policy can succeed are either where there is substantial agreement with the policy by direct producers (as was the case in South Korea), or where there is a massive security operation which militarily and strategically overwhelms any opposition (as in the case of the Republic of South Africa in some instances). The first is authoritarian but not coercive, the second is successfully coercive through an overwhelming means of oppression.

A strategic choice of conservation policy (and of other related issues such as land reform, pricing policy for agricultural inputs and outputs, and decentralisation of decision-making, revenue generation and disposal) must be aware of these elements in rural development, and the problems which they create. The strategic choice refers to an approach to erosion and conservation which is consistent with a broader view of development, and which recognises the problems with *all* current and past development models.

3. Practical pessimism

While the outlook for major success in conservation looks bleak, it is not uniformly so. One of the most inappropriate responses to the possibilities of successful conservation is a catatonic pessimism. There are directions to be followed which can help to bring about some success, even if short-lived or small-scale. Here the suggestions may seem tame and mildly reformist, but they are the only ones which are feasible, and it is perhaps better to end this book in an honest whisper than a spurious bang.

One objection to the view taken in this book that high rates of surplus extraction from peasants and pastoralists force them to extract non-renewable nutrients and energy from the soil, pasture and forests, is that there is no precise, quantitative and determining connection between cause and effect. It may be argued that it has to be demonstrated that a reduction in surplus extraction leads to a reduction in soil erosion. Indeed, there are cases when better prices for commodities benefit the direct producers. This has led to an ecologically unsound expansion of production onto marginal land. The reply to this objection demonstrates the complexity of the situation, and the impracticality of a model which posits a single causal variable. Exploitative relationships between direct producers and others are manifold, both in the purely economic, as well as political spheres. A reduction in surplus extraction implies many other changes as well if environmental conservation is to take place. First of all, a shared perception of the problem by land-users and institutions or arrangements of their choice to ensure a reasonably equitable sharing of costs and benefits of conservation are both essential – and these are not necessarily implied by a reduction of surplus extraction. Thus better prices for producers, reduction in rents, or security of share-croppers by themselves may not substantially reduce soil erosion. Second, there has to be the technical knowledge of conservation which may bring direct producers into close contact with officials, and the officials have to be able to prescribe appropriate conservation measures and be able to offer their services in such a way as to be utilisable by land-users. Third, there has to be a collective discipline in implementing conservation works and in policing. If *all* these conditions are met, then conservation has a good chance of success. Therefore a reduction in surplus extraction does not imply merely a small economic adjustment, but it must involve a transformation of the whole set of relationships between direct land-users and others.

A productive outcome of this approach to soil erosion and conservation is to focus upon the implications of erosion for growing inequality and impoverishment. Soil erosion is frequently both a result of and a contributory cause of inequality, and it should be quite feasible to undertake research to show what the impact will be upon the incomes of the most disadvantaged, as well as upon national economic statistics, particularly its imports of foodstuffs, timber and other forest products and its agricultural exports. Politicians, government ministers, senior bureaucrats and foreign consultants can use the results of this research in pointing out the wider contradictions between impoverished sectors of the population who both directly cause and suffer from soil erosion and the more advantaged who frequently maintain those inequalities – and reproduce the conditions for continued erosion. These political-economic analyses and critiques are bound to be embarrassing and unpopular.

Still within the realm of ideas, practical projects, even on a small scale, must be used to change people's minds about how they relate to each other and the environment. Social forestry in the strict sense of the word (forestry for local ends, benefiting local society), technologies for small peasants such as agro-silviculture, small cooperative ventures in soil conserving land use and erosion works, rascal-proof systems of local management and control of watersheds, forest land, fuelwood lots, water resources and such like – all of these types of project – can be used to demonstrate the vital link between the democratic control of all local land-users of their environment and successful conservation. These are the kinds of project which are very slow to show results, which need painstaking research and follow-up. Frequently they stand a better chance the less they have to do with central government. Local peasant organisations, rural trade unions, women's groups and other often fragmented and politically fragile institutions can be encouraged and financed by voluntary or non-governmental organisations, but small, fund-starved schemes have to withstand constant pressure from central government and local vested interests. From reading reports of such

156 The political economy of soil erosion

schemes, it seems that positive results are achievable but it takes a long time to build up the political acumen and technical knowledge to make a conservation programme stick. They cannot ever hope to become the 'answer' to conservation since they tend to run counter to all the powerful interests discussed in this book. While the conservation element may not be of any great importance to large landlords, senior government officials and so on, the political organisation of small peasants and even the landless which is implied will be of much greater concern – and it is this vital element which is necessary for successful conservation.

On the other hand, the various institutions of state can also have a progressive part to play, in that they have the resources to undertake useful research and development in agriculture. The fact that they seldom do in ways to help poor farmers and pastoralists living in marginal (and often eroded) areas, has already been discussed in Chapter 2. The political and economic signals from these people (and the crops they produce) will not be strong enough to divert resources and adapt resource and development techniques to the problems of conservation and reclamation. Again it is a question of swimming against the current, in this case against many agricultural scientists, their funding sources, and politicians and senior bureaucrats who have a direct responsibility and interest in the output of agricultural research stations. Research programmes involving intensive farmer involvement into soil conserving tillage practices (perhaps also involving enhanced nutritional value) can be carried out using adaptive research and outreach on a continuous and evolving basis, throughout the research programme. Cheap methods of reclaiming eroded hillsides (other than costly replant-ing), further research on agro-silviculture, nutrition-enhancing and soil conserving intercropped rotations, fuel-conserving improved cooking stoves, and other cheap energy-saving devices are all relatively unattractive 'maintenance research' which need to be pursued by an act of will. Acts of will can be likened to random perturbations of particles in a slowly moving draught of air – the general trend allows for them, and even counter-current movements over short periods. But if that is all that is feasible, it is better than nothing.

States, governments and bureaucracies are never monolithic. There are struggles for power, conflicts of interests, the able and the moribund, alliances and conspiracies and, in the case of soil conservation, often lassitude and inactivity. In the interstices of government, a soil conservationist can still successfully pursue policies which may run counter to most interests of persons in government and in official politics. Political alliances have to be made and colleagues chosen carefully.

The final contradiction still remains. This book on soil

conservation has been generally pessimistic about the future of the environment in lesser developed countries and the poorer people who use it. All it can realistically do is to call for a new intellectual approach and individual action – something which, by its own method, is deemed not to be enough.

Bibliography

Adams, B. (1982) 'Third World E.I.A. – a whole new ball game', *Ecos*, 3,3: 30–5.

Adamson, C.M., Melville, I.R. & Kanieki, G.T. (1982) *An Integrated Approach to Land Development and Soil Conservation in an Agricultural Settlement Project*. Soil and Water Conservation Workshop, Nairobi, Kenya, March 1982.

Agarawal, B. (1980) *The Woodfuel Problem and the Diffusion of Rural Innovations*. Science Policy Research Unit, University of Sussex.

Alavi, H. (1973) 'Peasant classes and primordial loyalties', *J. of Peasant Studies*, 1, 1: 23–54.

Albert, D. (1979) 'Hugging trees', *Win*, 9, 22.11.1979.

Alier, J.M. & Naredo, J.M. (1982) 'A Marxist precursor of energy economics: Podolinksy', *J. of Peasant Studies*, 9, 2: 207–23.

Allan, W. (1967) *The African Husbandman*. Oliver and Boyd, Edinburgh.

Allen, R. (1980) *How to Save the World. A Strategy for World Conservation*. Kogan Page, London.

Alvares, C. (1982) 'A new mystification?', *Development Forum*, Jan.–Feb., pp. 2–4.

Amin, S. (1974) *Accumulation on a World Scale; a Critique of the Theory of Underdevelopment*, trans. by Pearce, B. Monthly Review Press, New York.

Amin, S. (1977), *Imperialism and Unequal Development*. Hassocks, Harvester Press, New York.

Amos, M.J. (1982) 'Economics of Soil Conservation'. Unpublished MSc dissertation, National College of Agricultural Engineering, Silsoe, Bedfordshire.

Anon (1982), 'Planned activity note for the Mahiti Project', ST Bus Stand, Above Jain Market Street, Dhandhuka 382460, Ahmadabad District, India, 22 pps. (mimeo).

Arledge, J.E. (1980) 'Soil conservation at work: Guatemala's small farm project', *J. of Soil and Water Conservation*, 35 4: 187–9.

Ashworth, J.H. & Newendortter, J.W. (1982) 'Escaping the rural energy dilemma: a process for matching technologies to local needs and resources', *World Development*, 10, 4: 305–18.

Aweto, A.O. (1981) 'Fallowing and soil fertility restoration in South West Nigeria', *J. of Tropical Geography*, 3, June, 1–7.

Bagchi, A.K. (1982) *The Political Economy of Underdevelopment*. Modern Cambridge Economics Series, Cambridge University Press.

Bajracharya, D. (1981) *Implications of Fuel and Food Needs for Deforestation: an Energy Study of a Hill Village Panchayat of Eastern Nepal*.

Unpublished PhD dissertation, University of Sussex.

Baker, M.L. (1980) 'National Parks, conservation and agrarian reform in Peru', *Geographical Review*, **70**, 1: 1–18.

Baker, P.R. (1980) *Desertification: Cause and Control*. School of Development Studies Occasional Paper No. 6, May 1980, University of East Anglia, Norwich.

Baker, P.R. (1981) 'Environmental degradation in Kenya: Social Crisis or Environmental Crisis?' Paper for Developing Areas Study Group of the Institute of British Geographers Conference, Leicester, January 1981. Also DEV Discussion Paper No. 82, School of Development Studies, University of East Anglia, Norwich.

Baldwin, M. (1954) 'Soil erosion survey of Latin America', *J. of Soil and Water Conservation*, **9**, 7: 158–68.

Bali, J.S. (1974) *Soil Degradation and Conservation Problems in India*. New Delhi, Ministry of Agriculture, Government of India.

Banerji, D. (1977) 'Family planning programme', *Economic and Political Weekly*, XII, 6/7/8: 261–5.

Banister, J. & Thapa, S. (1981) *The Population Dynamics of Nepal*. Papers of the East-West Population Institute, No. 78.

Barbira-Scazzocchio, F. (ed) (1980) *Land, People and Planning in Contemporary Amazonia*. Centre of Latin American Studies Occasional Publication No. 3, Cambridge University.

Barlowe, R. (1958) *Land Resource Economics*. Prentice-Hall, Englewood Cliffs.

Bartlett, H.H. (1956) 'Fire, primitive agriculture and grazing in the tropics', in Thomas, W.L. (ed.). *Man's Role in Changing the Face of the Earth*, University of Chicago Press, Chicago, Illinois.

Bauer, P.T. (1976) *Dissent on Development*. Weidenfeld and Nicholson, London (also 1972, Harvard University Press, Cambridge, Mass.).

Beck, L. (1981) 'Government policy and pastoral land use in South West Iran', *J. of Arid Environments*, **4**, 253–67.

Beckerman, W. (1974) *In Defence of Economic Growth*. Jonathan Cape, London.

Beets, W.C. (1982) *Multiple Cropping and Tropical Farming Systems*. Gower, Westview Press, London.

Bein, F.L. (1980) 'Response to drought in the Sahel', *J. of Soil and Water Conservation*, **35**, 3: 121–4.

Beinhart, W. (1981) '*Development' Policy and Rural Resistance in South African Reserves 1930-1960*. Queen Elizabeth House, Oxford.

Belshaw, D.G.R. (1979) 'Taking indigenous technology seriously: the case of inter-cropping techniques in East Africa' *IDS Bulletin*, **10**, 2, 24–27.

Bennett, H.H. (1944) 'Food comes from the soil', *Geographical Review*, **34**, 57–76.

Bennett, H.H. (1955) *Elements of Soil Conservation*, 2nd edn. McGraw Hill Book Co., Washington DC.

Benton, T. (1977) *Philosophical Foundations of the Three Sociologies*. Routledge & Kegan Paul, London.

Bernal, J.D. (1969) *Science in History*, 4 vols. (4th edn). Penguin, Harmondsworth.

Bernard, F.E. & Thom, D.J. (1981) 'Population pressures and human carrying capacity in selected locations of the Machakos and Kitui districts', *J. of Developing Areas*, **15**, 3: 381–406.

Bernstein, H. (1977) 'Notes on capital and peasantry', *Review of African Political Economy*, **10**, 64–5.

Bernstein, H. (1979) 'African peasantries: a theoretical framework', *J. of Peasant Studies*, **6**, 4: 420–44.

Berry, L. & Townsend, J. (1973) 'Soil conservation practices in Tanzania: an historical perspective' in Rapp A., Berry L., & Temple, P. (eds.), *Studies of Soil Erosion and Sedimentation in Tanzania*, Research Monograph No. 1, BRALUP, University of Dar es Salaam, Tanzania.

Biggs, S.D. (1981) 'Agricultural research: a review of social science analysis', Report to the International Development Research Center, Ottawa. Also School of Development Studies Discussion Paper No. 115, University of East Anglia, Norwich.

Biggs, S.D. (1982a) 'Population and technology'. Book review for *Third World Quarterly*, **4**, 3.

Biggs, S.D. (1982b) 'Generating agricultural technology: Triticale for the Himalayan Hills', *Food Policy*, **7**, 1: 69–82.

Biggs S.D. (1983) 'Monitoring and control in agricultural research systems: maize in north India', *Research Policy*, **12**, 1: 37–59.

Blaikie, P.M. (1970) 'Spatial organisation of agriculture in some north Indian villages: Part I', *Transactions of the Institute of British Geographers*, **52**, 1–40.

Blaikie, P.M. (1971) 'Spatial organisation of agriculture in some north Indian villages: Part II', *Transactions of the Institute of British Geographers*, **53**, 15–30.

Blaikie, P.M. (1972) 'Implications for selective feedback in aspects of family planning research for policy makers in India', *Population Studies*, **26**, 3: 437–44.

Blaikie, P.M. (1975) *Family Planning in India: Diffusion and Policy*. Edward Arnold, London.

Blaikie, P.M. (1979) 'Poor peasants', Chapter 3 in *Peasants and Workers in Nepal*, edited by Seddon, J.D., Blaikie, P.M. & Cameron, J. Aris & Phillips, Warminster.

Blaikie, P.M. (1981) 'Class, land-use and soil erosion', *ODI Review*, **2**, 57–77. Also in *The Political Economy of Development and Underdevelopment*, Wilber, C.K. (ed.) (forthcoming).

Blaikie, P.M. (1983a) 'The political economy of soil erosion' in O'Riordan, T. & Turner, R.K. (eds.) *Progress in Resource Management and Environmental Planning*, **4**, 29–55. John Wiley & Sons, Chichester.

Blaikie, P.M. (1983b) 'How can soil erosion be a political matter?' *Pokhara Review*, **1**, 1: 25–39, Pokhara, Nepal.

Blaikie, P.M., Cameron, J., Fleming, R. & Seddon, J.D. (1977), *Centre, Periphery and Access: Social and spatial relations of inequality in west-central Nepal*. Development Studies Monograph No. 5, School of Development Studies, University of East Anglia, Norwich, 146 pp. (mimeo).

Blaikie, P.M., Cameron, J. & Seddon, J.D. (1979) *The Struggle for Basic Need: a Case Study of Nepal*. OECD Development Centre, Paris.

Blaikie, P.M., Cameron, J. & Seddon, J.D. (1980) *Nepal in Crisis: Growth and Stagnation at the Periphery.* Oxford University Press, London.

Blaikie, P.M., Cameron, J. & Seddon, J.D. (1981) 'The logic of a basic needs strategy: with or against the tide?' Paper presented at a Seminar on Basic Needs Strategy as a Planning Parameter, German Foundation for International Development, Berlin. Also DEV Discussion Paper No. 79, School of Development Studies, University of East Anglia, Norwich.

Blandford, D.C. (1981) 'Rangelands and soil erosion research' in Morgan, R.P.C. (ed), *Soil Conservation: Problems and Prospects*, John Wiley, Chichester, pp. 105–21.

Blasco, M. (1979) 'La tierra en el desarrollo rural da la Zona Andina', *Desarrollo Rural en las Americas*, **11**, 3: 155–65.

Bo Mycong Woo (1982) 'Soil erosion control in South Korea', *J. of Soil and Water Conservation*, **37**, 3: 149–50.

Bonsu, M. (1981) 'Assessment of erosion under different cultural practices on a savanna soil in the northern region of Ghana' in Morgan, R.P.C. (ed.), *Soil Conservation: Problems and Prospects*, John Wiley & Sons, Chichester, pp. 247–53.

Booth, A. & McCawley, P. (1981) *The Indonesian Economy during the Soeharto Era*, Oxford University Press, Kuala Lumpur & London.

Bose, A. (1978) 'The Sanjay factor', *India Today*, 1–15 Feb., pp. 40–1.

Boserup, E. (1965) *The Conditions of Agricultural Growth.* Allen & Unwin Ltd, London.

Boserup, E. (1981) *Population and Technology.* University of Chicago, Chicago, Illinois.

Bovill, E.W. (1920) 'The encroachment of the Sahara on the Sudan', *J. of the African Society*, **XX**, 175–85.

Bradby, B. (1975) 'The destruction of natural economy', *Economy and Society*, **4**, 2: 127–61.

Bradley, P.N. (1981) *Development Research and Physical Geography: Issues of Bias and Relevance.* Department of Geography, University of Newcastle-upon-Tyne, Newcastle, 9 pp. (mimeo).

Brandon, J.B. & Ghicoine, D.L. (1982) *An Evaluation of USAID's Soil and Water Management Project in Gambia.* Department of Agricultural Economy, University of Illinois, 17 pp. (mimeo).

Brandt, V. (1978) 'The new community movement: planned and unplanned change in rural Korea', *J. of Asian and African Studies*, *XIII*, **3**, 4: 196–211.

Brandt, W. (ed.) (1980) *North-South: A Programme for Survival. The Report of the Independent Commission on International Development Issues.* Pan Books, London & Sydney.

Braun, A. (1977) 'A collective discipline', *Ceres*, No. 69, 12, 3: 19–24.

Brett, E.A. (1973) *Colonialism and Underdevelopment in East Africa: the Politics of Economic Change 1919-1939.* Heinemann, London.

Brink R.A., Densmore, J.W. & Hill, G.A. (1977) 'Soil deterioration and the growing world demand for food', *Science*, **197**, 4304: 625–30.

Briscoe, J. (1979) 'Energy use and social structure in a Bangladesh village', *Population and Development Review*, **5**, 4: 615–41.

Brokensha, D., Warren, D. & Werner, O. (eds.) (1980) *Indigenous*

Knowledge Systems and Development, University Press of America, Washington D.C.

Brookfield, H.D. (1981) 'Man, environment and development in the outer islands of Fiji', *Ambio*, **X**, 2–3: 59–67.

Brookfield, H.D. (1982) *Intensification and Disintensification Revisited.* Department of Human Geography, Research School of Pacific Studies, Australian National University, Canberra, 33 pp. (mimeo).

Brow, J. (1976) *Population, Land and Structural Change in Sri Lanka and Thailand.* E.J. Brill, Leiden, Netherlands.

Brown, L. (1969) *Seeds of Change: the Green Revolution and Development in the 1970s.* Overseas Development Council, Praeger, New York.

Brown, L. (1973) *Population and Affluence: Growing Pressure on World Food Resources.* Overseas Development Council, Paper 15.

Brown, L. (1979) 'Where has all the soil gone?', *Mazingira*, **10**, 61–8.

Brown, L. (1981) 'Eroding the basis of civilisation', *J. of Soil and Water Conservation*, **36**, 5: 255–60.

Bunting, A.H. (1978) 'Science and technology for human needs: rural development and the relief of poverty', OECD Workshop on Scientific and Technological Cooperation with Developing Countries, Paris, 11.4. 1978.

Bunyard, P. (1980) 'Terraced agriculture in the Middle-East', *The Ecologist*, **10**, 8–9: 312–16.

Burbach R. & Flynn P. (1980) *Agribusiness in the Americas.* Monthly Review Press, New York & London.

Burke, A.E. (1956) 'Influence of man upon nature – the Russian view', in Thomas, W.L. (ed.), *Man's Role in Changing the Face of the Earth*, University of Chicago Press, Chicago, Illinois.

Bwalya, M.C. (1980) 'Rural differentiation and poverty reproduction in Northern Zambia: the case of Mpika District', in *The Evolving structure of Zambian Society.* Proceedings of a seminar held in the Centre for African Studies, Edinburgh University, Scotland, 295 pp. (mimeo).

Cahill, D.N. (1981) *Soil Conservation in India.* FAO, New Delhi.

Caldwell, D. (1978) *Food Crises and World Politics.* Sage Publications, London.

Caxner, G. (1982) 'Survival, interdependence and competition among the Philippine rural poor', *Asian Survey*, **22**,4, 369–84.

Cassen, R.H. (1976) 'Population and development: a survey', *World Development*, **4**, 10/11: 785–830.

Cassen, R.H. (1980) *India: Population, Economy, Society.* Macmillan, London.

Cater, L.J. (1977) 'Soil erosion: the problems persist despite the billions spent on it', *Science*, **196**, 4288: 409–11.

CEPAL (1976) *El Medio Ambreite en America Latina.* United Nations Economic and Social Council, E/CEPAL/1018.

Ceres 31 (1973) 'The decolonisation of agriculture in Algeria', **6**, 1: 24–8. (by Ollivier, M.).

Ceres (1982) 'Study to gauge environmental costs of reclamation', **14**, 4: 7 (author anonymous).

Centre of African Studies (1980) *The Evolving Structure of Zambian Society.*

Proceedings of a seminar held in the Centre of African Studies, University of Edinburgh, 30 and 31 May 1980, 295 pp. (mimeo).

Chambers, R. (1980a) *Who gets a last rural resource?: the potential and challenge of lift irrigation for the rural poor*. Institute of Development Studies, Sussex (mimeo).

Chambers, R. (1980b), 'Rural poverty unperceived: problems and remedies', *IBRD Working Paper*, No. 400, IBRD, Washington DC.

Chambers, R. (1983) *Putting the Last First: Priorities in Rural Development*. Longman, London.

Chambers, R. & Singer, H. (1980) *Poverty, Malnutrition and Food in Zambia*. Country Case Study for World Development Report IV, Institute of Development Studies, Sussex, Brighton.

Chauvin, H. (1981) 'When an African city runs out of fuel' *Unasylva*, **33**, 133: 11–20.

Cherenvisinov, G.A. (1964) 'Eroded soils of the forest-steppe zone in the European part of the USSR, agronomic characteristics and means of their efficient utilisation', 8th International Congress of Soil Science, Bucharest, Rumania, pp. 721–33.

Choudhury, K. (1982a) 'Agro-forestry: the rural poor and institutional structures', Workshop on Agro-forestry (United Nations University and Albert-ludwigs-Universität, Freiburg), 31 May – 5 June 1982.

Choudhury, K. (1982b) 'The greening of India', *The Indian Express*, 21 Dec. 1982, New Delhi, India.

CIMMYT (1980) *Planning Technologies Appropriate to Farmers: Concepts and Procedures*. Centro Internacional de Mejoramiento de Maiz y Trigo, Mexico City, Mexico.

Clayton, E. (1964) *Agrarian Development in Peasant Economies*. Pergamon Press, Oxford.

Cleaver, H.M. (1972) 'The contradictions of the Green Revolution', *Monthly Review*, **24**, 80–111.

Cliffe, L. (1964) 'Nationalisation and the reaction to agricultural improvement during the enforced colonial period', Paper to the EAISR conference, Makerere University, Uganda, December 1964.

Cliffe, L. (1978) 'Labour, migration and peasant differentiation: the Zambian experience', *J. of Peasant Studies*, **5**, 3: 326–46; also in Turok, B. (ed.), *Development in Zambia: A Reader* (1981), Zed Press, London.

Cliffe, L. & Moorsom, R. (1979) 'Rural class formation and ecological collapse in Botswana', *Review of African Political Economy*, **15–16**: 35–52.

Cloudsley-Thompson, J.L. (1973) 'The expanding Sahara', *Environment Conservation*, **1**, 1: 5–13.

Coale, A.J. & Hoover, E.M. (1958) *Population Growth and Economic Development in Low Income Countries*. Princeton University Press, Princeton.

Cohen, B.J. (1973) *The Question of Imperialism: the Political Economy of Dominance and Dependence*. Basic Books Inc., New York.

Cohen, R., Gutkind, P.C.W. & Brazier, P. (eds.) (1979) *Peasants and Proletarians: The Struggles of Third World Workers*. Hutchinson, London.

Colsen, E. (1971) *The Social Consequences of Resettlement: the Impact of the Kariba Resettlement upon the Gwembe Tonga*. (Institute of African Studies, University of Zambia). Manchester University Press, Manchester.

Comte, M.C. (1980) 'Making social forestry work', *Ceres*, **74**, 41–4.

Conway, G. & Romm, J. (1973) *Ecology and Resource Development in South East Asia*. Report to Ford Foundation (mimeo).

Conyers, O. (1971) *Agro-economic Zones in Tanzania*. BRALUP Research Department, Dar es Salaam, Tanzania.

Cook, K. (1982) 'Soil loss: a question of values', *J. of Soil and Water Conservation*, **37**, 2: 89–92.

Cook, K. (1983) 'Surplus madness', *J. of Soil and Water Conservation*, **38**, 1: 25–8.

Copans, J. (1975) *Qui le nourrit de la famine en Afrique?* and *Sécheresses et famines du Sahel* (2 vols.). F. Maspero, Paris.

Cottrell, A. (1977) *Environmental Economics*. Edward Arnold, London.

Coulson, A. (1975) 'Peasants and bureaucrats', *Review of African Political Economy*, **3**, May–Oct., 53–8.

Coulson, A. (1981) 'Agricultural policies in mainland Tanzania, 1946-76' in Heyer J., Roberts P. & Williams, G. (eds.), *Rural Development in Tropical Africa*, Macmillan, London.

Courier (1980) 'Against the grain', May 1980, pp. 10–14.

Crossan, P. & Miranowski, J. (1982) 'Soil protection: why, by whom, and for whom?', *J. of Soil and Water Conservation*, **37**, 1: 27–9.

Crowe, B.L. (1969) 'The tragedy of the commons re-visited', *Science*, **116**, 1103–7. Also reprinted in, Hardin, G.J. & Baden, J. (eds.), *Managing the Commons* (1977), W.H. Freeman, San Francisco, California.

Curtis, R.U. (1979) *Subtropical Lands' Development*. USAID/BOLIVIA, Washington DC. Evaluation of AID project, 511-T-050.

Datoo, A. (1977) 'Peasant agricultural production in East Africa: menace and consequences of dependence', *Antipode*, **9**, 1: 70–5.

de Bandt J.D., Mandi, P. & Seers, D. (1980) *New Trends in European Development Studies*. European Studies in Development, Macmillan, London.

Deere, C.D. & de Janvry, A. (1979), 'A conceptual framework for the empirical analysis of peasants', *American Journal of Agricultural Economy*, November, pp. 601–11.

de Janvry, A. (1981) *The Agrarian Question and Reformism in Latin America*. John Hopkins University Press, Baltimore, and London.

Delfs, R. (1982) 'The petrified forests', *Far Eastern Economic Review*, 20 Aug. 1982, p. 61.

Delfs, R. (1982) 'The price of neglect', *Far Eastern Economic Review*, 20 Aug. 1982, pp. 62.

Denevan, W. (1980) 'Traditional agricultural resource management in Latin America' in Klee, G.A. (ed.), *World Systems of Traditional Resource Management*, Winston & Sons Press, New York.

Dequi, J., Leidi, Q. & Jusheng, T. (1981), 'Soil erosion and causation in the Winding River Valley, China' in Morgan, R.P.C. (ed.), *Soil Conservation: Problems and Prospects*, John Wiley & Sons, Chichester.

Deutsch, K.W. (ed.) (1977) *Eco-social Systems and Eco-politics*. UNESCO, Paris.

de Wilde, J.C. (1967) *Experiences with Agricultural Development in Tropical Africa*, Vol. 1, 2. John Hopkins University Press, Baltimore.

Dhogra, B. (1980) 'Victims of ecological ruin', *The Ecologist*, 10, 5: 170–2.

Digerness, T.H. (1979) 'Fuelwood crisis causing unfortunate land use – and the other way round', *Norsk Geogr. Tidsskr*, 33, 23–32.

Dinham, B. & Hines, C. (1983) *Agribusiness in Africa*. Earth Resource Research Ltd., London.

Djurfeldt, G. & Lindberg, S. (1975), *Behind Poverty: The Social Formation in a Tamil Village*. Scandinavian Institute of Asian Studies Monograph Series, No. 22. Oxford & IBH Publishing Co., New Delhi, India.

Doran, M.H., Low, A.R.C. & Kemp, R.L. (1979) 'Cattle as a store of wealth in Swaziland: implications for livestock development and over grazing in Eastern and Southern Africa', *Annual Journal of African Economic Development*, 61, 1: 41–7.

Douglas, I. (1981) 'Soil conservation measures in river basin planning' in Saha, S.K. & Barrow, C.J. (eds.), *River Basin Planning: Theory and Practice*, John Wiley & Sons, Chichester.

Douglas, J.J. (1982) 'Traditional fuel usage and the rural poor in Bangladesh', *World Development*, 10, 8: 669–76.

Dudal, R. (1981) 'An evaluation of conservation needs', in Morgan, R.P.C. (ed.), *Soil Conservation: Problems and Prospects*, John Wiley & Sons, Chichester.

Dumont, R. (1970) *Types of Rural Economy: Studies in World Agriculture*. Methuen, London.

Dumsday, R. (1971) 'Evaluation of soil conservation policies by systems analysis' in Dent, J.B. & Anderson, J.R. (eds.), *Systems Analysis in Agricultural Management*, John Wiley & Sons, Chichester.

Dumsday, R. & Flinn, J. (1977) 'Evaluating systems of soil conservation through bio-economic modelling' in Greenland, O. & Lal, R. (eds.), *Soil Conservation and Management in the Humid Tropics*, John Wiley & Sons, Chichester.

Earl, O.E. (1975) *Forest Energy and Economic Development*. Clarendon Press, Oxford.

Eckholm, E.P. (1976a) 'The politics of soil conservation', *The Ecologist*, 6, 2: 54–9.

Eckholm, E.P. (1976b) 'The other energy crisis: firewood', *The Ecologist*, 6, 3: 80–6.

Eckholm, E.P. (1978) *Losing Ground: Environmental Stress and World Food Problems*. Pergamon Press, London.

Eckholm, E.P. (1982) *Down to Earth: Environment and Human Needs* (International Institute for Environment and Development). Pluto Press, London.

Eco-Systems (1982) *Southeast Shinyanga Land Use Study*, Vol. 1, Survey of Land Utilisation; Vol. 3, Planning for Soil Conservation. Reports to Shinyanga Rural Integrated Development Programme, Tanzania & IBRD, Nairobi.

Ehrlich, P.R., Ehrlich, A.H. & Holdren, J.P. (1973) *Ecoscience: Population,*

Resources, Environment. W.H. Freeman, San Francisco, California.

Elliot, R.H. (1888) 'Indian famines', *Proceedings of the Royal Colonial Institute*, **IX**, 1887-8: 2–41.

El-Swaify, S.A., Dangler, E.W. & Armstrong, C.L. (1982) *Soil Erosion by Water in the Tropics*. College of Tropical Agriculture and Human Resources, University of Hawaii, Hawaii.

Elvin, M. (1973) *The Pattern of the Chinese Past*. Eyre Methuen, London.

England, R. & Bluestone, B. (1972) 'Ecology and social conflict' in Daly, H.E. (ed.), *Towards a Steady State*, W.H. Freeman, San Fransisco, California, pp. 190–214.

Ensenberger, H.M. (1974) 'A critique of political ecology', *New Left Review*, **8**, 4: 3–32.

Ervin, O.E. & Washburn, R.A. (1981) 'Profitability of soil conservation practices in Missouri', *J. of Soil and Water Conservation*, **36**, 4

ESCAP (1979) *The Relationship of Migration and Urbanisation to Economic Development in Korea*, Bangkok, Thailand.

Ethiopian Soil Conservation Workshop Participants (1982) 'Soil conservation in Ethiopia: a brief look at the history, problems and achievements', National College of Agricultural Engineering, Silsoe, England (July-Aug. 1982).

FAO (1960) *Soil Erosion by Wind*. FAO, Rome.

FAO (1965) *Soil Erosion by Water*. FAO, Rome.

FAO (1966) *African Agricultural Development*. FAO, Rome.

FAO (1971) *Shifting Cultivation in Latin America*. FAO Forestry Development Paper No. 17 by R.F. Watters, FAO, Rome.

FAO (1976a) *Desert Creep and Range Management in the Near East*. Mission Report, March-April 1976, FAO, Rome.

FAO (1976b) *Soil Conservation for Developing Countries*. FAO Soils Bulletin No. 30, FAO, Rome.

FAO (1977) *Soil Conservation and Management in Developing Countries*. FAO Soils Bulletin No. 33, FAO, Rome.

FAO (1977b) *Assessing Soil Degradation*. FAO Soils Bulletin No. 34, FAO, Rome.

FAO (1978a) *Methodology for Assessing Soil Degradation*. Report on FAO/UNEP Expert Consultation, FAO, Rome.

FAO (1978b) *The State of Food and Agriculture 1977*. FAO Agricultural Series No. 8, FAO, Rome.

FAO (1979) *A Provisional Methodology for Soil Degradation Assessment*. FAO, Rome.

FAO (1980) *Natural Resources and the Human Environment for Food and Agriculture*. Environment Paper No. 1, FAO, Rome.

FAO (1982a) *Appropriate Technology in Forestry*. SIDA/FAO Report on Intermediate Technology in Forestry, New Delhi, 18 Oct. – 7 Nov. 1981, FAO, Rome.

FAO (1982b) *Tropical Forest Resources*, by J.P. Lanly, FAO, Rome.

FAO (1982c) *Village Forestry Development in the Republic of Korea: a Case Study*. GCP/INT/347/SWE, FAO, Rome.

Feder, E. (1977) 'Agribusiness and the elimination of the Latin American rural proletariat', *World Development*, **5**, 5–7, May/July.

Feldman, R. (1975) 'Rural social differentiation and political goals in

Tanzania', in Oxaal, I., Barnett, T. & Booth, D. (eds.), *Beyond the Sociology of Development*, Routledge & Kegan Paul, London.

Fernea, R.A. (1969) 'Land reform and ecology in post-revolutionary Iraq', *Economic Development and Cultural Change*, **17**, 3: 356–79.

Feuchtwang, S. (1973) 'The discipline and its sponsors' in Assad, T. (ed.), *Anthropology and the Colonial Encounter*, Ithaca House, New York, pp. 71–102.

Finer, S.E. (1976) *The Man on Horseback: The Role of the Military in Politics*. Peregrine Books, London.

Floret, C. & Hadjaj, M.S. (1977) 'An attempt to combat desertification in Tunisia', *Ambio*, **6**, 6: 366–7.

Floyd, B.N. (1972) 'Land Apportionment in Southern Rhodesia', in Mansell Prothero, R. (ed.), *People and Land in Africa South of the Sahara*, Oxford University Press, Oxford.

Fontaine, R. (1981) 'What is really happening to tropical forests?', *Ceres*, No. 82, 14, 4: 15–19.

Fosbrooke, H. & Young, R. (1976) *Land and Politics among the Luguru of Tanganyka*. Routledge & Kegan Paul, London.

Frank, A.G. (1967) *Capitalism and Underdevelopment in Latin America: Historical Studies in Chile and Brazil*. Monthly Review Press, New York.

Frank, A.G. (1980) *Crisis: in the Third World*. Heinemann, London.

Franke, R. & Chasin, B.H. (1980) *Seeds of Famine: Ecological Destruction and the Development Dilemma in the Western Sahel*. Allenheld, Osmun & Co., New Jersey.

Franke, R. & Chasin, B.H. (1981) 'Peasants, peanuts, profits and pastoralists', *The Ecologist*, **11**, 4: 156–68.

Freeman, P. (1980) *Bolivia: State of the Environment and Natural Resources*. Contract for ASAID, by JRB Associates, Virginia, USA.

French, D. (1982) 'The Ten Commandments of renewable energy analysis', *World Development*, **10**, 1: 71–9.

Friedmann, J. & Weaver, C. (1979) *Territory and Function*. Edward Arnold, London.

Gaige, F.H. (1975) *Regionalism and National Unity in Nepal*. University of California Press, Berkeley, California.

Gallopin, G. & Berrera, C. (1979) 'The nexus society and environment' in Gallopin, G. (ed.), *Environment and Styles of Development: Some Conceptual and Methodological Issues*, Technical Research Project No. 35, IFDA.

Gardner, N. (1981) 'Small-scale Soil Degradation Mapping'. Dept. of Geography, University of Reading (mimeo).

George, S. (1976) *How the Other Half Dies*. Penguin, Harmondsworth.

Gilbert, A. (1974) *Latin American Development – a Geographical Perspective*. Penguin, Harmondsworth.

Gilbert, E.H., Norman, D.W. & Winch, F.E. (1980) 'Farming systems research: a critical appraisal', *MSU Rural Development Paper No. 6*, Dep. of Agricultural Economics, Michigan State University, Michigan.

Glacken, C.J. (1956) 'Changing ideas of the habitable world' in Thomas, W.J.F. (ed.), *Man's Role in Changing the Face of the Earth*, University of Chicago Press, Chicago.

Glantz, M.H. (ed.) (1977) *Desertification*. Westview Special Studies in Natural Resources and Energy Management, Westview Press, Boulder, Colorado.

Glover, Sir H. (1946) *Erosion in the Punjab: its Causes and Cure*. Feroz Printing Works, Lahore, Pakistan.

Goldman, M. (1972) 'Ecological facelifting in the USSR or improving upon nature' in I. Sachs (ed.), *The Political Economy of Environment: Problems of Method*. Mouton, Paris.

Goldsmith, E. (1980) 'WEAP – World Ecological Areas Programme – a proposal to save the world's tropical rain forests', *The Ecologist*, **10**, 1: 6–52.

Goodman, D.E. (1977) 'Rural structure, surplus mobilisation and modes of production in a peripheral region: the Brazilian North East', *J. of Peasant Studies*, **5**, 1: 3–32.

Goudie, A. (1981) *The Human Impact, Man's Role in Environmental Change*. Basil Blackwell, London, U.K.

Gould, P.R. (1969) 'Man against his environment: a game theoretical framework' in Vayda, A.P. (ed.), *Environment and Cultural Behaviour, Ecological Studies in Cultural Anthropology*, The Natural History Press, New York.

Grainger, A. (1980) 'The state of the world's tropical rain forests', *The Ecologist*, **10**, 1: 6–52.

Greenland, D. & Lal, R. (eds.) (1977) *Soil Conservation and Management in the Hunid Tropics*. John Wiley & Sons, Chichester.

Gregory, D. (1978) *Ideology, Science amd Human Geography*. Hutchinson, London, U.K.

Gribbin, J. (1982) 'The other face of development', *New Scientist*, **24** (Nov. 25): 489–95.

Griffin, K. (1972) *The Green Revolution: an Economic Analysis*. United Nations Research Institute for Social Development, Geneva, Switzerland.

Griffin, K. (1979) *The Political Economy of Agrarian Change (Second Edition 1980)*. Macmillan, London.

Grigg, D.B. (1970) *The Harsh Lands*. Macmillan, London.

Grigg, D.B. (1980) *Population Growth and Agrarian Change*. Cambridge University Press.

Grunwald, J. & Musgrove, P. (1970) *Natural Resources in Latin American Development*. Johns Hopkins Press, Baltimore.

Gustafsson, J.E. (undated 1981?) *Land Utilisation and Agricultural Development in the Peoples Republic of China since 1949*, Part 1 (mimeo). International Centre for Soil Conservation Information, Silsoe, Bedfordshire.

Gwatkin, D.R. (1979) 'Political will and family planning: the implications of India's Emergency experience', *Population and Development Review*, **5**, 1: 29–59.

Halcrow, H.G., Heady, E.O. & Cotner, M.L. (eds.) (1982) *Soil Conservation, Policies, Institutions and Incentives*. Published by Soil Conservation Society of America for North Central Research Committee 11, Iowa.

Hardin, G.J. (1972) 'The tragedy of the commons' in Daly, G.H.E. (ed.), *Towards a Steady-State Economy*, W.H. Freeman, San Francisco, California, pp. 133–48; also in *Science*, **162**, 1243–8.

Hardin, G.J. (1977) *The Limits to Altruism: an Ecologist's View of Survival.* Bloomington, Indiana.

Hardin, G.J. & Baden, J. (eds.) (1977) *Managing the Commons.* W.H. Freeman, San Francisco, California.

Harmon, L., Knutson, R. & Rosenberg, P. (1979) 'Soil depletion study reference report for southern Iowa Rivers Basin', Iowa Department of Agriculture, Soil Conservation Service, Iowa.

Harriss, J. (1980) 'Why poor people stay poor in rural south India', *Development and change*, **11**, 1: 33–64.

Harriss, J. (1982) *Capitalism and Peasant Farming: Agrarian Structure and Ideology in Northern Tamil Nadu.* Oxford University Press, Bombay, India.

Harshbarger, C.E. & Swanson, E.R. (1964) 'Soil loss tolerance and the economics of soil conservation in the Swygert hills', *Illinois Agricultural Economics*, **4**, 2: 18–28.

Hart, D.M. (1976) *The Ait Waryaghar of the Moroccan Rif: an Ethnography and History.* Viking Fund Publications No. 55, University of Arizona Press, Arizona.

Heathcote, R.L. (1980) *Perception of Desertification.* United Nations University, Tokyo, Japan.

Hedfors, L. (1982) *Evaluation and Economic Appraisal of Soil Conservation in a Pilot Area*, Ministry of Agriculture, Nairobi, Kenya (mimeo).

Heichelheim, F.M. (1956) 'Effects of classical antiquity on the land' in Thomas, W.L. (ed.), *Man's Role in Changing the Face of the Earth*, University of Chicago Press, Chicago, Illinois.

Hellen, J.A. (1977) 'Legislation and landscape: some aspects of agrarian reform and agricultural adjustment', in O'Keefe, P. & Wisner, B. (eds.), *Land Use and Development*, African Environment Special Report No. 5, International African Institute, London.

Herrera, A. (1977) 'The risks involved', *Mazingira*, **3/4**: 25–9.

Heusch, B. (1981) 'Sociological constraints on soil conservation: a case study, the Rif Mountains, Morocco' in Morgan, R.P.C. (ed.), *Soil Conservation: Problems and Prospects*, John Wiley & Sons, Chichester.

Heyer, J., Roberts, P. & Williams, G. (1981) *Rural Development in Tropical Africa.* Macmillan, London.

Hiraoka, M. & Yamamoto, S. (1980) 'Agricultural development in the Upper Amazon of Ecuador', *Geographical Review*, **70**, 4: 423–45.

Holdgate, M.W. (1982) 'The environmental information needs of the decision maker', *Nature and Resources*, **XVIII**, 1: 5–9.

Holdgate, M.W., Kassas, M. & White, G.F. (eds.) (1982) *The World Environment 1972-1982.* United Nations Environmental Programme, published by Tycooly International, Dublin, Ireland.

Holland, S. (1980) *The Regional Problem.* Macmillan, London.

Holliman, J. (1972) *The Ecology of World Development.* Voluntary Committee on Overseas Aid and Development, London.

Holtman, J.B. & Connor, J.L. (1979) 'Potential corn yield and related economic incentives for soil conservation', *Transactions of the*

American Society of Agricultural Engineers, **22**, 1: 75–80.

Hopkins, A. (1973) *An Economic History of West Africa*. Longman, London.

Howard, M.C. & King, J.E. (1975) *The Political Economy of Marx*. Longman, London.

Howard, P.H. (1981) 'Impressions of soil and water conservation in China', *J. of Soil and Water Conservation*, **36**, 3: 122–4.

Hudson, N. W. (1971) *Soil Conservation*. Batsford, London.

Hudson, N.W. (1981) 'Non-technical constraints on soil conservation' in Tingsanchati, T. and Eggers, H. (eds.), *South East Asian Regional Symposium on Problems of Soil Erosion and Sedimentation*, Bangkok, Thailand.

Hunt, D. (1973) 'Poverty and agricultural development policy in a semi-arid area of Eastern Kenya' in O'Keefe, P. and Wisner, B. (eds.), *Landuse and Development*, African Environment Special Report 5, International African Institute, London.

Hyams, E. (1952) *Soil and Civilisation*. 2nd Edn, John Murray, London (1976).

Hyden, G. (1980) *Beyond Ujamaa in Tanzania: Underdevelopment and an Uncaptured Peasantry*. Heinemann, London.

IBRD (1977a) *India's Population Policy: History and Future*. World Bank Staff Working Paper No. 265 (prepared by Gulhati, R.) Washington DC.

IBRD (1977b) *Forestry: Sector Policy Paper*. World Bank Report No. 1778, Washington DC.

IBRD (1979) *Nepal: Development Performance and Prospects*. World Bank Country Study, Washington DC.

ICRISAT (1979) 'Development and transfer of technology for rainfed agriculture and the SAT farmer', Proceedings of conference, Patancheru, India, 28 Aug. – 1st Sept.

Ives, B. (1981) 'Crisis in the Himalayas', *Development Forum* (September).

Jacks, G.V. & Whyte, R.O. (1939) *Vanishing Lands: a World Survey of Soil Erosion*. Doubleday Doran, New York.

Jarvis, L. & Klein, E. (1975) 'Employment creation and the conservation of natural resources: a programme for El Salvador', *Trabajos Ocasionales*. ILO, Programa Regional del Empleo para America latina y El Cariba.

Jodha, N.S. (1980) 'The process of desertification and the choice of interventions', *Economic and Political Weekly*, 9 Aug., pp. 1351–5.

Johnson, P. (1978) 'Fighting the frontier fever', *Ceres*, **64**, (July/Aug.): 23–30.

Kapp, K.W. (1972) 'The implementation of environmental policies', *Development and Environment*, United Nations, Geneva, Switzerland.

Kartawinata, K., Adisoemarto, S., Riswan, S. & Vayda, A. (1981) 'The impact of man on a tropical forest in Indonesia', *Ambio*, **10**, 2–3: 115–19.

Kassapu, S. (1979) 'The impact of alien technology', *Ceres*, No. 67, 12, 1: 29–33.

Kay, G. (1972) 'Resettlement and land-use planning in Zambia: the Chipangali scheme' in Mansell Prothero, R. (ed.), *People and Land in Africa South of the Sahara*, Oxford University Press, Oxford.

Kayastha, S.L. (1964) *The Himalayan Beas Basin: A Study in Habitat, Economy and Society*. Banaras Hindu University, Banaras, India.

Kellogg, C.E., (1941) *The Soils that Support Us*. Macmillan, New York.

Kennedy, B.A. (1979) 'A naughty world', *Transactions of the Institute of British Geographers, New Series*, 4, 550–8.

Khanbanonda, C. (1972) *Thailand's Public Law and Policy for Conservation and Protection of Land, with Special Attention to Forests and Natural Areas*. PhD Dissertation, Indiana University, USA.

Kilakuldilok, S. (1981) *Deforestation in Thailand: A Study of its Causes and the Role of the Government*. Dissertation, School of Development Studies, University of East Anglia, Norwich.

Kirby, E.S. (1972) 'Environmental spoilage in the USSR', *New Scientist*, 6 Jan. 1972.

Kirkby, M.J. (1980) 'Soil erosion in Britain: is it an enigma?' *Progress in Physical Geography*, 4, 24–27.

Kirkby, M.J. & Morgan, R.P.C. (eds.) (1980) *Soil Erosion*. John Wiley and Sons, Chichester.

Ki-Zerbo, J. (1981) 'Women and the energy crisis in the Sahel', *Unasylva*, 33, 33: 5–10.

Klepper, R. (1980) 'The state and peasantry in Zambia' in *The Evolving Structure of Zambian Society*. Proceedings of a Seminar, Centre of African Studies, University of Edinburgh, Edinburgh, pp. 120–50.

Klepper, R. (1981) 'Zambian agriculture, structure and performance' in Turok, B. (ed.), *Development in Zambia: a Reader*, Zed Press, London.

Kon Muang Nan (1978) 'The problem of hill tribe people', *Vanasru*, 36, 2, 8–12.

Komarov, B. (1981) *The Destruction of Nature in the Soviet Union*. Pluto Press, London.

Kulkarni, S. (1983) 'Towards a social forest policy', *Economic and Political Weekly*, 5 Feb., pp. 191–6.

Labib, T.M. (1981) 'Soil erosion and total denudation due to flash floods in the Egyptian desert', *J. of Arid Environments*, 4, 3: 191–202.

Lacoste, Y. (1974) 'Fundamental structures of mediaeval North African Society', *Economy and Society*, 3, 1: 1–17.

Ladejinsky, W. (1969) 'Green revolution in the Kosi area of Bihar: a field trip', *Economic and Political Weekly*, A147-A161, September. (Review of Agriculture).

Lanly, J.P. (1982) *Tropical Forest Resources*. FAO Forestry Paper No. 30, Rome, Italy.

Lappé, F.M., Collins, J. & Kinley, D. (1980) *Aid as Obstacle: 20 Questions about our Foreign Aid and the Hungry*. Institute of Food and Development Policy, USA.

LASA (1978) *Lesotho's Agriculture: a Review of Existing Information*. Lasa Research Report, No. 2.

Lean, G. (1981) 'Hug-a-tree-campaign could stop a disaster', *Observer*, 8.9.1981. World Report, p.6.

Leslie, A.J. (1980) 'Logging concessions: How to stop losing money', *Unasylva*, 32, 129: 2–7.

Lesotho, 2nd 5-year Development Plan 1975-76–1979-80, Vol. 1.

Leys, C. (1975) *Underdevelopment in Kenya*, Heinemann, London.

Lima, L.C. (1971) 'When 7 percent is better than 9', *Ceres*, No. 23, 4, 5: 35–6.

Lindqvist, S. (1979) *Land and Power in South America*. Penguin Books, Harmondsworth.

Lipton, M. (1977) *Why Poor People Stay Poor: Urban Bias in World Development*. Temple Smith, London.

Lipton, M. (1982) 'Why poor people stay poor', in Harriss, J. (ed.), *Rural Development: Theories of Peasant Economy and Agrarian Change*. Hutchinson, London, pp. 66–81.

Llaurado, J.P. (1980) 'Forestry and agrarian reform', in *Proceedings of the Latin American Conference on the Conservation of Renewable Natural Resources*. IUCN 1968, published with the assistance of UNESCO.

Low, F. (1967) 'Estimating potential erosion in developing countries', *J. of Soil and Water Conservation*, 22, 4: 147–8.

Luxemburg, R. (1963) *The Accumulation of Capital*. Routledge & Kegan Paul, London (1964, Monthly Review Press, New York).

MAB (1979) *Trends in Research and the Application of Science and Technology for the Development of Arid Zones. Man and the Biosphere*, MAB Technical Notes No. 10, UNESCO, Paris.

MAB (1980) *Population-Environment Relations in Tropical Islands: the Case of Eastern Fiji*. Brookfield, H.C. (ed.), MAB Technical Notes No. 13, UNESCO, Paris.

Mabbutt, G.J.A. (1978) 'The impact of desertification as related by mapping', *Environment Conservation*, 5, 1: 45–56.

MacAndrews, C. & Sien, C.L. (eds.) (1979) *Developing Economies and the Environment, the Southeast Asian Experience*. McGraw-Hill, Washington DC.

Malcolm, D.W. (1938) Sukumuland – an African People and their Country. International African Institute and Oxford University Press, Oxford.

Mamdani, M. (1973) *The Myth of Population Control: Family Caste and Class in an Indian Village*. Monthly Review Press, New York.

Mamdani, M. (1976) 'The ideology of population control', *Economic and Political Weekly*, **XI**, 31–3, Special Number, 1143–6.

Mansell-Prothero, R. (ed.), (1972) 'Some observations on desiccation in North-Western Nigeria', in *People and Land in Africa, South of the Sahara*, Oxford University Press, Oxford.

Marter, A. & Honeybone, D. (1976) *The Economic Resources of Rural Households and the Distribution of Agricultural Development*. Rural Development Studies Bureau, University of Zambia, Lusaka, Zambia (mimeo).

McCowan, R.L., Haaland, G. & de Haan, C. (1979) 'The interaction between cultivators and livestock producers in semi-arid Africa' in Hall, A.E., Cannell, G.H. & Lawton, H.W. (eds.), *Agriculture in Semi-Arid Environments*, Springer-Verlag, Heidelberg.

Meadows, D.H., Meadows, D.L. & Anders, J. (1972) *The Limits to Growth*, Earth Island, London, U.K.

Mesarovic, M.D. & Pestel, E. (1976) *Mankind at the Turning Point*, Dutton, New York.

Metcalf, W.J. (1977) *The Environmental Crisis: A Systems Approach*.

University of Queensland Press, Australia.

Mikesell, M. (1969) 'The deforestation of Mount Lebanon', *Geographical Review*, **59**, 1–28.

Miliband, R. (1969) *The State in Capitalist Society*. Weidenfeld, London; and Quartet Books (1973), London.

Misiko, P.A.M. (1982) 'Strategies and constraints and planning of soil and water management programmes on the catchment basis in Western Kenya', presented at Soil and Water Conservation Workshop, Faculty of Agriculture, University of Nairobi, March 1982.

Mitra, S.K. (1982) 'Ecology as science and science fiction', *Economic and Political Weekly*, **XVII**, 5: 147–52.

Mnzava, E.M. (1981) 'Fuelwood: the private energy crisis of the poor', *Ceres* No. 82, 14, 4: 35–9.

Morgan, R.P.C. (1979) *Soil Erosion*, Topics in Applied Geography Longman, London.

Morgan, R.P.C. (ed.) (1981) *Soil Conservation: Problems and Prospects*. John Wiley & Sons, Chichester.

Murray, G.F. (1980) 'Haitian peasant contour ridges: the evolution of indigenous erosion control technology', Development Discussion Paper No. 86, Harvard Institute for International Development, Cambridge, USA, 54 pp.

Myers, N. (1979) *The Sinking Ark*. Pergamon Press, Oxford.

Nagrobrahman, G.D. & Sambrani, S. (1983) 'Women's drudgery in firewood collection', *Economic and Political Weekly*, **XVIII**, 1 and 2: 33–8.

Nobe, G.K. & Seckler, D.W. (1979) *An Economic and Policy Analysis of Soil Water Problems in the Kingdom of Lesotho*. Lasa Research Report No. 3, Ministry of Agriculture, Maseru Lesotho.

Oasa, E.K. & Jennings, B.H. (1982) 'Science and authority in international agricultural research', *Bulletin of Concerned Asian Scholars*, **14**, 4: 30–44.

ODG (1981) *Agriplan Training System: Case Studies and Development of Agriculture in Central Province, Zambia*. Report to the Food and Agriculture Organisation, Rome. Overseas Development Group, Norwich.

Odum, H.T. (1971) *Environment, Power and Society*. John Wiley & Sons, Chichester.

Ohlin, G. (1969) 'Population pressure and alternative investments' in IUSSP International Population Conference, Vol. 3, London.

O'Keefe, P. (1975) *African Drought: A Review*. Disaster Research Unit, University of Bradford, Occasional Paper No. 8, Bradford.

O'Keefe, P. Wisner, B. & Baird, A. (eds.) (1977), in *Kenyan Underdevelopment: A Case Study of Proletarianisation*, African Environment Special Report No. 5, International African Institute, London.

Okigbo, B.N. (1981) 'Alternatives to shifting cultivation'. *Ceres*, **84**, 6: 41–5.

Olayide, S.O. & Falusi, A.O. (1977) 'Economics of soil conservation and erosion control practices in Nigeria' in Greenland D. and Lal, R. (eds.), *Soil Conservation and Management in the Humid Tropics*, John Wiley & Sons, Chichester.

Onweluzo, B.S.K. & Onyemelukwe, J.O.C. (1977) 'Forest influences on environemtal stability', in O'Keefe, P. & Wisner, S. (eds.), *Landuse and Development*, African Environment Special Report No. 5, pp. 18–32, International African Institute, London.

Optima (1981), C-210 Defence Colony, N. Delhi, 1100224, *Forestry India 1981: an historical and current sector review.* P.G. Ramachadran (SIDI).

Palmer, R. & Parsons, N. (1977) *The Roots of Rural Poverty in Central and Southern Africa.* Heinemann, London.

Parikh, J.K. (1977) 'Environmental problems of India and their possible trends', *Environmental Conservation*, **4**, 3.

Patnaik, N. (1975) 'Soil erosion: a menace to the (Indian) nation', *Indian Farming*, **24**, 11: 7–11.

Payer, C. (1982) *The World Bank: a Critical Analysis.* Monthly Review Press, New York and London.

Pearse, A. (1980) *Seeds of Plenty, Seeds of Want: Social and Economic Implications of the Green Revolution.* United Nations Research Institute for Social Development, Geneva, Switzerland. Clarendon Press, Oxford.

Pearson, C. & Pryor, A. (1978) *Environment: North and South – An Economic Interpretation.* John Wiley, New York.

Pearson, L.B. (ed.) (1964) *Partners in Development: Report on the Commission for International Development.* Pall Mall Press, London.

Pickering, K. (1979) 'Soil conservation and rural institutions in Java', *IDS Bulletin*, June 1979, **10**, 4. Guest Issues, DEV Studies, Birmingham.

Pimentel, D., Terhune, E.C., Dyson-Hudson, R., Rochereau, S., Samis, R., Smith, E.A., Denham, D., Reifschneider, D. Strepard, and M. (1976) 'Land degradation effects on food and energy resources', *Science*, **194**, 4261: 149–55.

Plumwood, V. & Routley, R. (1982) 'World rainforest destruction – the social factors' *The Ecologist*, **12**, 1: 4–22.

Posner, J.L. & McPherson, M.F. (1981) *The Steep-sloped Areas of Tropical America: Current Situation and Prospects for the Year 2000.* Agricultural Sciences Division, Rockefeller Foundation, New York (mimeo). 21 pp. plus bibliography.

Posner, J.L. & MacPherson, M.F. (1982) 'Agriculture on the steep slopes of tropical America: the current situation and prospects', *World Development*, May, 341–54.

Preston, D. (ed.) (1980) *Environment, Society and Rural Change in Latin America.* John Wiley & Sons, Chichester.

Pryde, P.R. (1972) *Conservation in the Soviet Union.* Cambridge University Press, Cambridge.

Qu Geping (1980) 'Deserts in China and their prevention and control', *Mazingira*, **4**, 2: 74–9.

Quick, S.A. (1977), 'Bureaucracy and rural socialism in Zambia', *J. of Modern African Studies*, **15**, 3: 379–400.

Rafiq, M. (1978) *The Present Situation and Potential Hazard of Soil Degradation in 10 Countries of the Near East Region.* FAO, Rome.

Raikes, P. (1977) 'Rural differentiation and class formation in Tanzania', *J. of Peasant Studies*, **5**, 3: 285–319.

Raikes, P. & Coulson, A. (1975) 'Ujaman and rural socialism', *Review of*

African Political Economy, May-Oct. **3**, 53–8.

Ranganathan, S. (1978) 'A plan for rural India', *The Ecologist*, **8**, 1: 10–12.

Ranger, T.O. (1971) *The Agricultural History of Zambia*. National Educational Company of Zambia, Lusaka, Zambia, 28 pp.

Ranger, T.O. (1978) 'Growing from the roots: reflections on peasant research in Central Southern Africa', *J. of Southern African Studies*, **5**, 1: 99–133.

Rapp, A. (1975) 'Soil erosion and sedimentation in Tanzania and Lesotho', *Ambio*, **4**, 4: 154–63.

Rapp, A., Berry, L. & Temple, P. (1973) *Studies of Soil Erosion and Sedimentation in Tanzania*. Research Monograph No. 1, Bureau of Resource Assessment and Land-use Planning, University of Dar es Salaam, Tanzania.

Rapp, A., le Houerou, H.N. & Loudholm, B. (eds.) (1976) 'Can desert encroachment be stopped?', *Ecological Bulletins*, NFR/24 pub. UNEP and SIES, Stockholm.

Rasmussen, A. (1982) 'Commentary', *J. of Soil and Water Conservation*, **37**, 1:24.

Rauschkolb, R.S. (1971) 'Land degradation', *FAO Soils Bulletin*, No. 3, Rome, Italy.

Redclift, M. (1982) *Poor Environments of Environmental Poverty? Towards a Political Economy of Natural Resource Use*. Paper delivered to the Annual Conference of Development Studies Association, Dublin, Ireland, 23–25th Sept. 17 pp. (mimeo).

Rennie, D.A. (1982) 'The deteriorating soils of the Canadian Prairie', *Span*, **25**, 3: 99–101.

Rennie, K. (1978) Quoted in Ranger, T.O. *Southern African Studies*, **5**, 1: 113.

Rey, L. (1959) 'Persia in perspective–2', *New Left Review*, 20, 69–72.

Richards, P. (1975) '"Alternative" strategies for the African environment. "Folk ecology" as a basis for community orientated agricultural development' in Richards, P. (ed.), *African Environment, Problems and Perspectives*, African Environmental Report No. 1, International African Institute, London.

Richards, P. (1979) 'Community environmental knowledge in African rural development', IDS Bulletin, **10**, 2, 28–36.

Rieger, H.C. (1978/79a) 'Socio-economic aspects of environmental degradation in the Himalayas', *J. of the Nepal Research Centre*, **2/3** (Sciences), 177–84.

Rieger, H.C. (1978/79b) 'An approach to a dynamic ecosystems model for a watershed', *J. of the Nepal Research Centre*, **2/3** (Sciences), 111–29.

Riquier, J. (1978) 'A methodology for assessing soil degradation', *Background Paper No. 1 presented to FAO/UNEP*. Expert Consultation and Methodology for Assessing Soil Degradation, FAO, Rome, Jan. 1978.

Riquier, J. (1982) 'A world assessment of soil degradation', *Nature and Resources*, **XVIII**, 2, Apr.–June: 18–21.

Roberts, A. (1979) *The Self-Managing Environment*. Allison and Busby, London.

Robinson, D.A. (1978a) *A Critical Review of Soil Conservation Policies and* their Implementation in Zambia 1940-1974. University of Sussex, Research Paper in Geography, Sussex.

Robinson, D.A. (1978b) *Soil Erosion and Soil Conservation in Zambia*. Occasional Study No. 9, University of Zambia, Lusaka.

Robinson, P. (1977) *The Environment Crisis: a Communist View*. The Communist Party Headquarters, London.

Roche, L. (1978) The practice of Agri-silviculture in the Tropics with special reference to Nigeria, FAO Soils Bulletin, No. 24. Rome.

Roder, W. (1977) *Environmental Assessment of the Rural Development Area Program, Swaziland*. University of Cincinnati and University of Zambia.

Rodney, W. (1972) *How Europe Underdeveloped Africa*. Bogle-L'Ouverture Publications, London.

Rogers, E.M. & Shoemaker, F.F. (1971) *Communication of Innovations: a Cross-Cultural Approach*. 2nd Edn, Free Press, New York.

Rounce, N.V. (1949) *The Agriculture of the Cultivation Steppe of the Lake, Western and Central Provinces*. Longman, Capetown, Republic of South Africa.

Routley, R. & Routley, V. (1980) 'Destructive forestry in Melanesia and Australia', *The Ecologist*, **10**, 1: 56–7.

Rowland, J.W. (1974) 'The Conservation ideal – Southern African regional commission for the conservation and utilisation of the soil', in Barnet, R.J. (ed.), *The Lean Years: Politics in the Age of Scarcity*, Abacus, Pretoria, Republic of South Africa.

Ruttan, V.W. (1981) 'An induced innovation interpretation of technical change in agriculture of development countries', Seminar *Cambio Tecnico en la Agro Latinoamericano: Situacion y Perspectivas en la Decada de 1980*. IICA/PNUD, Coronado, Costa Rica.

Ruttan, V.W. (1982) *Agricultural Research Policy*. Department of Agricultural and Applied Economics, University of Minnesota, USA.

Saha, S.K. (1979), 'River basin planning in the Damodar Valley of India', *Geographical Review*, **69**, 3: 273–87.

Saith, A. & Tankha, A. (1972) 'Agrarian transition and the differentiation of the peasantry. A study of a West V.P. village', *Economic and Political Weekly*.

Sandbach, F. (1980) *Environment, Ideology and Policy*. Basil Blackwell, Oxford.

Schertz, D.L. (1983) '*The basis for soil loss tolerances*', Journal of Soil & Water Conservation, **38**, 1, 10–14.

Schlich, P. (1889-90) 'Forestry in the colonies and in India' in *Proceedings of the Royal Colonial Institute*, **XXI**, 187–238.

Schmithüsen, F. (1979) 'Logging and legislation', *Unasylva*, **31**, 124: 2–10.

Schroeder, R. (1977) *Ecological Change in Rural Nepal: the Case of Batulechaur*. PhD dessertation, University of Washington, Washington DC (University Microfilms 1978).

Seddon, J.D. (1981) *Moroccan Peasants: a Century of Change in the Eastern Rif*. Dawson & Sons, London.

Sen, A. (1981) *Poverty and Famines: an Essay on Entitlement and Deprivation*. Oxford University Press, Oxford.

Sepuldeva, S. (1980) 'The effects of modern technologies on income distribution: a case of integrated rural development in Columbia', *Desarrollo Rural en las Americas*, **12**, 2: 105–25.

Shanin, T. (1973) 'The nature and logic of the peasant economy: a generalisation', *J. of Peasant Studies*, **1**, 1: 63–80.

Shaxson, T.F. (1981) 'Reconciling social and technical needs in conservation work on village farmlands' in Morgan, R.P.C. (ed.), *Soil Conservation: Problems and Prospects*, John Wiley & Sons, Chichester, pp. 385–97.

Sheddick, V. (1954) *Land Tenure in Basutoland.* HMSO, London.

Sheng, T. & Michaelson, T. (1973) 'Run-off and soil loss studies in yellow yams: forest development and watershed management in the upland regions', Jamaica FO:SF/JAM 505 Project Working Document, Kingston, Jamaica.

Simon, J.L. (1981) *The Ultimate Resource*, Martin Robertson, Oxford, and University of Princeton Press, Princeton.

Simons, H.J. (1981) 'Zambia's urban situation' in Turok, B. (ed.), *Development in Zambia: a Reader*, Zed Press, London.

Sinha, R. (1977) 'Agribusiness: a nuisance in every respect?', *Mazingira*, **3/4**, 16–23.

Skidmore, E.L. (1977) 'Criteria for assessing wind erosion' in *Assessing Soil Degradation*. FAO Soils Bulletin No. 34, pp. 52–62, Rome, Italy.

Skidmore, E.L., Fisher, P.S. & Woodruff, N.P. (1970) 'Wind erosion equation: a computer solution and application', *Soil Science Society Proceedings*, **34**, 931–5.

Smil, V. (1979) 'Controlling the Yellow River', *Geographical Review*, **69**, 3: 251–72.

Sorbo, G.M. (1977) 'Nomads on the scheme – a study of irrigation, agriculture and pastoralism in Eastern Sudan' in O'Keefe, P. & Wisner, B. (eds.), *Kenyan Underdevelopment: A Case Study of Proletarianisation*, African Environment Special Report No 5, International African Institute, London.

Spears, J.S. (1980) 'Can farming and forestry coexist in the tropics?' *Unasylva*, **32**, 128: 2–12.

Spears, J.S. (1982) 'Preserving watershed environments', *Unasylva*, **34**, 137: 10–14.

Spitz, P. (undated) *Droughts, Reserves and Social Classes (with special reference to India).* United Nations Research Institute for Social Development, Geneva, Switzerland, 43 pp. (mimeo).

State of India's Environment: a Citizen's Report, Centre for Science and Environment, New Delhi.

Stebbing, E.P. (1935) 'The encroaching Sahara', *Geographical Journal*, **85**, 6: 506–24.

Stewart, P. (1970) 'Erosion, a social problem', *Ceres*, **15**, 3: 3.

Stewart, P.J. (1975) *Algerian Peasantry at the Crossroads: Fight Erosion or Migrate.* A case study of Rural Development Employment. IDS Discussion Paper No. 69, Institute of Development Studies, University of Sussex, Brighton.

Stocking, M.A. (1978a) 'Prediction and estimation of erosion in sub-tropical Africa', *Geo-Eco-Trop*, **2**, 161–74.

Stocking, M.A. (1978b) 'Relation of agricultural history and settlement to severe soil erosion in Rhodesia', *Zambesia*, **VI**, 2: 129–45.

Stocking, M.A. (1981a) 'Conservation strategies for less developed countries' in Morgan, R.P.C. (ed.), *Soil Conservation: Problems and Prospects*, John Wiley & Sons, Chichester, pp. 377–83.

Stocking, M.A. (1981b) *Farming and Environmental Degradation in Zambia: the Human Dimension*. School of Development Studies, University of East Anglia, Norwich.

Stocking, M.A. (1982) 'Reasons to be bullish about Tanzania's Erosion', *International Agricultural Development*, pp. 8–9.

Stocking, M.A. (1983) *Development Projects for the small Farmer, Lessons from East and Central Africa in Adapting Soil Conservation*, Soil Conservation Society of America Conference Proceedings, Paper No.105. Also Overseas Development Group, University of East Anglia, Norwich.

Stocking, M.A. & Ellwell, H.A. (1976) 'Rainfall erosivity over Rhodesia', *Transactions of the Institute of British Geographers* (New Series), **1**, 2: 231–45.

Stocking, M.A. & Pain, A. (1983) '*Soil life and the minimum soil depth for productive yields: developing a new concept*'. Discussion Paper No. 150, School of Development Studies, University of East Anglia, Norwich.

Suharso, & Speare, A. (1981) 'Migration trends' in Booth, A. & McCawley, P. (eds.), *The Indonesian Economy during the Soeharto Era*, Oxford University Press, Kuala Lumpur & Oxford.

Swift, J. (1977a) 'Desertification and Man in the Sahel' in O'Keefe, P. & Wisner, B. (eds.), *Kenyan Underdevelopment: A Case Study of Proletarianisation*, African Environment Special Report No. 5, International African Institute, London.

Swift, J. (1977b) 'Pastoral development in Somalia: herding co-operatives as a strategy against desertification and famine' in Glantz, M.H. (ed.), *Desertification: Environmental Degradation in and around Arid Lands*, Westview Press, Boulder, Colorado.

Swift, J. (1979) 'Notes on traditional knowledge, modern knowledge and rural development'; IDS Bulletin, **10**, 2, 41–43.

Szeftel, M. (1980) 'The political process in post-colonial Zambia: the structural basis of factional conflict' in *The Evolving Structure of Zambian Society*. Proceedings of a Seminar in the Centre of African Studies, Edinburgh, Scotland, pp. 64–95.

Tapper, R. (1979) *Pasture and Politics*. Academic Press, London.

Taussig, M. (1978) 'Peasant economics and the development of capitalist agriculture in the Cauca Valley, Columbia', *Latin America Perspectives*, **18**, 5, 62–98

Tempany, W.H. & Grist, D.H. (1958) *An Introduction to Tropical Agriculture*. Longman, London.

Temple, P. (1973) 'Soil and water conservation policies in the Uluguru mountains, Tanzania' in Rapp, A., Berry, L. & Temple, P. (eds.), *Studies of Soil Erosion and Sedimentation in Tanzania*, Research Monograph Number 1, Bureau of Resource Assessment and Land-Use Planning, University of Dar es Salaam, Tanzania.

Temple, P. & Rapp, A. (1973) 'Landslides in the Mgera area, Western Uluguru mountains, Tanzania', in Rapp, A., Berry, L., & Temple,

P. (eds.), *Studies of Soil Erosion and Sedimentation in Tanzania.* Research Monograph No. 1, Bureau of Resource Assessment and Land-Use Planning, University of Dar es Salaam, Tanzania.

Tennakoon, M.U.A. (1980) 'Desertification in the dry zone of Sri Lanka' in Heathcote, R.L. (ed.), *Perception of Desertification*, United Nations University, Tokyo, Japan.

Terray, E. (1974) 'Long distance exchange and formation of the state: the case of the Abron Kingdom of Gyaman', *Economy and Society*, **3** 315–45.

Thomas, W. (Jr) (1956) 'Discussion: subsistence economies' in *Man's Role in Changing the Face of the Earth.* University of Chicago Press, Chicago, Illinois.

Thompson, H. & Warburton, M. (undated, 1982?) *Inter-relations Between Resources, Environment, People and Development: the Case of the Himalayan Foothills.* Discussion paper, International Institute for Applied Systems Analysis, Laxenburg, Austria, 93 pp. (mimeo).

Thomson, J.T. (1977) 'Ecological deteriorations: local level rule-making and enforcement problems in Niger', in Glantz, M.H. (ed.), *Desertification and Environmental Degradation in and around the Arid Lands*, Westview Press, Boulder, Colorado.

Thrupp, A. (1981) 'The peasant view of conservation', *Ceres*, No. 82, 14, 4: 31–4.

Timmons, J.F. (1980) 'Protecting agriculture's natural resource base', *Journal of Soil and Water Conservation*, **35**, 1, 5–11.

Tregubov, P.S. (1981) 'Effective erosion control in the USSR' in Morgan, R.P.C. (ed.), *Soil Conservation: Problems and Prospects*, John Wiley & Sons, Chichester, pp. 451–9.

Turner, B.L. (1974) 'Prehistoric intensive agriculture in the Mayan lowlands', *Science*, **185**, 31–4.

Turok, B. (ed.) (1981) *Development in Zambia: a Reader.* Zed Press, London.

UN (1977) *Desertification: its causes and consequences.* Background documents of the United Nations Conference on Desertification, Nairobi. Pergamon, Oxford.

UNEP (1979) *Industry and Environment.* Report No. 4, Nairobi, Kenya.

UNEP (1982) *The World Environment 1972-82*, Holdgate, M.W., Kassass, M. & White, G.F. (eds.), Natural Resources and the Environment Series, Vol. 8, published by Tycooly International, Dublin, Eire.

UNEP/FAO (1978) *The Present Situation and Potential Hazard of Soil degradation in Ten Countries of the Near East Region* (Rafiq, M.), FAO, Rome.

UNESCO (1979a) *Environmental Education of Engineers: Current Trends and Perspectives.* UNESCO, Paris.

UNESCO (1979b), *International Co-ordinating Council of the Programme on Man and the Biosphere.* UNESCO, Sixth Session, Paris.

UNESCO (1979c) *Trends in Research and the Application of Science and Technology for Arid Zone Development.* Man and the Biosphere, Technical Notes 10. UNESCO, Paris.

Unger, J. (1978) 'Collective incentives in the Chinese countryside: Lessons from Chen village', *World Development*, **6**, 583–601.

180   *The political economy of soil erosion*

USAID (1979a) *Draft Environmental Report on Nepal.* Science and Technology Division, Library of Congress, Washington DC.

USAID (1979b) *Draft Environmental Report on Thailand.* Science and Technology Division, Library of Congress, Washington DC.

USAID (1980) *Draft Environmental Profile on Swaziland.* Science and Technology Division, Library of Congress, Washington DC.

USAID (1980) *Bolivia: State of the Environment and Natural Resources* (Freeman, P. et al). JRB Associates, Virginia, USA.

USAID (1981) *Draft Environmental Profile on Rwanda.* Arid Lands Information Centre, Tucson, Arizona.

Valentin, C. & Roose, E.J. (1981) 'Soil and water conservation: problems in pineapple plantations of south Ivory Coast', in Morgan, R.P.C. (ed.), *Soil Conservation: Problems and Prospects*, John Wiley & Sons, Chichester, pp. 239–46.

Van Raay, H.G.T. & Lugo, A.E. (eds.) (1974) *Man and Environment Ltd.* Rotterdam University Press, Rotterdam.

Veblen, T.T. (1978), 'Forest preservation in the western highlands of Guatemala', *Geographical Review*, **68**, 4: 417–34.

Velloso, J.M. (1981) 'Who cares for the forest?' *Ceres*, No. 82, 14, 4: 40–3.

Viertmann, W. (1978) *Project Planning and Implementation in Mountain Watersheds, Northern Morocco.* Technical Report No. 3, FAO, Agricultural Department, Rome.

Vitebsky, P. (1980) *Political Relations among the Sora of India.* Paper I: Internal Relations, Paper II: External Relations, Girton College, University of Cambridge, Cambridge, (mimeo).

Wah, K.K. (1982) 'Progress and pollution', *New Internationalist*, No. 114.

Wallerstein, I. (1974) 'The rise and future demise of our world capitalist system: concepts for comparative analysis', *Comparative Studies in Society and History*, **16**, 387–415.

Ward, B. (1976) *The Home of Man.* Penguin, Harmondsworth.

Warren, B. (1980) *Imperialism: Pioneer of Capitalism.* New Left Books, London.

Warwick, D.P. (1982) *Bitter Pills.* Cambridge University Press, London and New York.

Watson, J.R. (1973) 'Conservation problems, policies and the origins of the Mlalo basin rehabilitation scheme, Usurubaru mountains, Tanzania' in Rapp, A., Berry, L., & Temple, P. (eds.), *Studies of Soil Erosion and Sedimentation in Tanzania*, Research Monograph No. 1, Bureau of Resource Assessment and Land-Use Planning, University of Dar es Salaam, Tanzania.

Watson, W. (1964) *Tribal Cohesion in a Money Economy: A Study of the Mambwe people of Northern Rhodesia.* Manchester University Press, Manchester.

Watt, K. E. F., Molloy, L. F., Varshney, C. K., Weeks, D. & Wirosardjono, S. (1977) *The Unsteady State: Environmental Problems, Growth and Culture.* East-West Centre, Honolulu.

Watts, D. (1968) 'Origins of Barbadian cane hole agriculture', *Barbados Museum and Historical Society Journal*, **32**, 143–51.

Weaver, P. (1979) 'Agro-silviculture in tropical America', *Unasylva*, **31**, 126: 2–11.

Webster, R.G. (1881-2) 'England's colonial granaries', *Proceedings of the Royal Colonial Institute*, **XIII**, 27.

Webster, C.C. & Wilson, P.N. (1980) *Agriculture in the Tropics*. Longman, London.

Weissman, S. (1970) 'Why the population bomb is a Rockefeller baby', *Ramparts*, May, pp. 86–91.

White, B. (1976) 'The economic importance of children in a Javanese village' in Nag, M. (ed.), *Population and Social Organisation*, Mouton, The Hague, Netherlands.

White, S. (1978) 'Cedar and mahogany logging in Eastern Peru', *Geographical Review*, **68**, 4: 394–416.

Whitlow, J.R. (1980) 'Environmental constraints and population pressures in the tribal areas of Zimbabwe', *Zimbabwe Agricultural Journal*, **77**, 3: 173–81.

Wiersum, K.F. (1980) 'Erosion, rural development and forests in Java/Erosie platteland Sontwikkeling en bos op Java', *Landboukundig Tijdschrift*, **92**, 9: 338–45.

Wiggins, S.L. (1981) 'The economics of soil conservation in the Acelhuate river basin, El Salvador' in Morgan, R.P.C. (ed.), *Soil Conservation: Problems and Prospects*, John Wiley & Sons, Chichester, pp. 339–415.

Wilde, J.C., assisted by McLoughlin, P.F.M., Guinard, A., Scudder, T. & Maubouché, R. (1967) *Experiences with Agricultural Development in Tropical Africa*, Vol. 2. Johns Hopkins Press, Baltimore.

Wilken, G.C. (1974) 'Some aspects of resource management by traditional farmers', in Biggs, H.H. & Tinnermeer, R.L. (eds.), *Small Farm Agricultural Development Problems*, Fort Collins, Colorado State University, Colorado.

Williams, G. (1976) 'Taking the part of peasants: rural development in Nigeria and Tanzania' in Gutkind, P. & Wallerstein, I. (eds.), *The Political Economy of Contemporary Africa*, Sage Publications, London.

Williams, G. (1981) 'The World Bank and the peasant problem' in Hayer, J., Roberts P. and Williams, G. (eds.), *Rural Development in Tropical Africa*, Macmillan, London.

Williams, W.H. (1893), 'Uganda', *Proceedings of the Royal Colonial Institute*, Vol. **XXV**, 109.

Winch, P. (1958) *The Idea of a Social Science and its Relation to Philosophy*. Routledge & Kegan Paul, London.

Wischmeier, W.H. (1976) 'Use and misuse of the universal soil loss equation', *J. of Soil and Water Conservation*, **31**, 5–9.

Wisner, B. (1976) 'Man-made famine in Eastern Kenya: the inter-relationship of environment and development', Institute of Development Studies Discussion Paper No. 96, Sussex.

Wisner, B. (1978) 'Does radical geography lack an approach to environmental relations?', *Antipode*, **10**, 1: 84–95.

Wisner, B., O'Keefe, P. & Westgate, K. (1977) 'Global systems and local disasters: the untapped power of people's science', *Disasters*, **1**, 1: 47–57.

Wolf, E. (1959) *Sons of the Shaking Earth: The People of Mexico and Guatemala – their land, history and culture*. University of Chicago, Chicago, Illinois.

Woo, B.M. (1982) 'Soil erosion control in South Korea', *J. of Soil and Water Conservation*, **37**, 3: 149–50.

Wood, G.D. (1976) 'Rural development and the post colonial state', University of Bath, U.K., 20 pp. (mimeo).

Young, A.T. (1960) 'Soil movement by denudational processes on slopes', *Nature*, **188**, 118–23.

Young, A. T. (1976) *Tropical Soils and Soils Survey*. Cambridge Geographical Studies No. 9, Cambridge University Press, Cambridge.

Young, R. & Fosbrooke, H. (1960) *Land and Politics among the Luguru of Tanganyika*. Routledge & Kegan Paul, London.

Zorina, Ye. F., Kosov, B.F. & Prokhorova, S.D. (1977) 'The role of the human factor in the development of gullying in the steppe and wooded steppe of the European USSR', *Soviet Geography*, Jan. **XVIII**, 1: 48–55.

Zumer-Linder, M. (1972) Preliminary summaries of national reports submitted by the developing countries to the Stockholm Conference on the Human Environment. June 1972, Stockholm.

Index